Modality, morality, and belief are among the most controversial topics in philosophy today, and few philosophers have shaped these debates as deeply as Ruth Barcan Marcus. Inspired by her work, a distinguished group of philosophers explore these issues, refine and sharpen arguments, and develop new positions on such topics as possible worlds, moral dilemmas, essentialism, and the explanation of actions by beliefs. This ''state-of-the-art'' collection honors one of the most rigorous and iconoclastic of philosophical pioneers.

AF722017

MODALITY, MORALITY, AND BELIEF

Modality, Morality, and Belief

Essays in Honor of Ruth Barcan Marcus

Edited by

WALTER SINNOTT-ARMSTRONG
Dartmouth College

In collaboration with

DIANA RAFFMAN
Ohio State University

NICHOLAS ASHER
University of Texas at Austin

CAMBRIDGE
UNIVERSITY PRESS

CAMBRIDGE UNIVERSITY PRESS
Cambridge, New York, Melbourne, Madrid, Cape Town, Singapore, São Paulo, Delhi

Cambridge University Press
The Edinburgh Building, Cambridge CB2 8RU, UK

Published in the United States of America by Cambridge University Press, New York

www.cambridge.org
Information on this title: www.cambridge.org/9780521440820

© Cambridge University Press 1995

© David Kaplan for Chapter 4

This publication is in copyright. Subject to statutory exception
and to the provisions of relevant collective licensing agreements,
no reproduction of any part may take place without the written
permission of Cambridge University Press.

First published 1995
This digitally printed version 2008

A catalogue record for this publication is available from the British Library

Library of Congress Cataloguing in Publication data

Modality, morality, and belief : essays in honor of Ruth Barcan Marcus
/ Walter Sinnott-Armstrong, Diana Raffman, Nicholas Asher [editors].

p. cm.

ISBN 0-521-44082-3

1. Modality (Logic) 2. Ethics, Modern – 20th century. 3. Belief
and doubt. 4. Marcus, Ruth Barcan. I. Marcus, Ruth Barcan.
II. Sinnott-Armstrong, Walter, 1955– . III. Raffman, Diana.
IV. Asher, Nicholas.
BC199.M6M62 1995 93-38811
160–dc20 CIP

ISBN 978-0-521-44082-0 hardback
ISBN 978-0-521-10057-1 paperback

Contents

Preface

Ruth Barcan Marcus is, and always has been, an iconoclast. Again and again she has shattered conventional wisdom, forcing us to look at old problems in new ways. Like any intellectual pioneer, she has often faced an onslaught of skepticism, but a brief survey of the current philosophical literature, including the essays published here for the first time, testifies to the enduring influence of her views.

Marcus is probably best known for her seminal work in philosophical logic. In 1946, she published the first systematic treatment of quantified modal logic, therein turning aside Quine's famous attack on the coherence of combining quantifiers with alethic operators. She later extended the first-order formalization to second order with identity (1947) and to modalized set theory (1963). Marcus's writings in logic either inaugurated or brought to the fore a number of issues that have loomed large in subsequent philosophical theorizing. Of special importance are the axiom known as the Barcan formula (1946), the theorem about the necessity of identity (1963), a flexible notion of extensionality (1960, 1961), and the idea that ordinary proper names are contentless directly referential tags (1961). This last laid the groundwork for the theory of direct reference later elaborated by Kripke, Donnellan, Kaplan, and others.

No less a revolutionary in moral theory, Marcus undermined the entire structure of standard deontic logic in her paper titled "Iterated Deontic Modalities" (1966). She later (1980) argued against some theorists that moral dilemmas are real, and against others that moral dilemmas need neither derive from inconsistent rules nor imply moral antirealism.

More recently, in a series of papers on belief (1981, 1983, 1990), Marcus has repudiated theories that identify beliefs with attitudes to linguistic or quasilinguistic items (sentences of English or "Mentalese," for example). She claims instead that for an agent *A* to believe that *p* is for *A* to be disposed to behave as if *p* obtains (where *p* is a possible state of affairs). Unlike its "language-centered" competitors, Marcus's analysis applies straightforwardly to the beliefs of nonlingual and infralingual systems. Perhaps most important, it mobilizes a conception of rational agents as seeking to maintain global coherence among the verbal and nonverbal indicators of their beliefs.

The essays published here were inspired by Marcus's work. Those in Part I concern themes from her writings on modality; those in Part II criticize or develop material in her papers on morality; and those in Part III treat issues addressed in her work on belief. Taken together, these essays illustrate the span and depth of her influence.

Marcus has been honored by many prestigious appointments and awards. Among other things, she was Professor of Philosophy and Chair of the Department at the University of Illinois at Chicago, Professor of Philosophy at Northwestern University, and, until the time of her retirement in 1992, Reuben Post Halleck Professor of Philosophy at Yale University. She has been a fellow of the Guggenheim Foundation, the Rockefeller Foundation, the Stanford Center for Advanced Study in the Behavioral Sciences, the Edinburgh Institute for Advanced Studies in the Humanities, the Mellon Foundation, the National Science Foundation, and the American Academy of Arts and Sciences. In 1986, she was awarded the medal of the Collège de France.

In addition to her scholarly contributions, Marcus's tireless service to the international philosophical community has been widely recognized. To cite just a few examples, she was elected Chairman of the National Board of Officers of the American Philosophical Association, President of the Association for Symbolic Logic, and President of the Institute International de Philosophie. She has also served on the editorial boards of numerous prominent journals.

Ever ready to engage in dialectic on virtually any topic, insistent upon absolute rigor in philosophical argument, and unfailingly generous with her time and support, Marcus has had a profound influence on generations of students. The editors of this volume have been privileged to have Ruth as a teacher and as a friend. We look forward to learning still more from her in the future. It is to her, with esteem and affection, that this volume is dedicated.

Bibliography of the Philosophical Works of Ruth Barcan Marcus

(Philosophical publications exclusive of abstracts and reviews, except where the review has had influence on subsequent literature)

BOOKS

The Logical Enterprise, ed. with A. R. Anderson and R. M. Martin. New Haven, Conn.: Yale University Press, 1975.

Logic, Methodology, and Philosophy of Science. Vol. VII, ed. R. B. Marcus, G. J. W. Dorn, and P. Weingartner. Amsterdam and New York: North-Holland, 1986.

Modalities: Philosophical Essays. New York: Oxford University Press, 1993.

ARTICLES

Published under "Ruth C. Barcan"

"A Functional Calculus of First Order Based on Strict Implication." *Journal of Symbolic Logic* 11 (1946).

"Deduction Theorem in a Functional Calculus of First Order Based on Strict Implication." *Journal of Symbolic Logic* 11 (1946).

"The Identity of Individuals in a Strict Functional Calculus of First Order." *Journal of Symbolic Logic* 12 (1947).

"Critical Review of Arthur Smullyan 'Modality and Description'." *Journal of Symbolic Logic* 13 (1948).

Published under "Ruth Barcan Marcus"

"The Elimination of Contextually Defined Predicates in a Modal System." *Journal of Symbolic Logic* 15 (1950).

"Strict Implication, Deducibility, and the Deduction Theorem." *Journal of Symbolic Logic* 18 (1953).

"Extensionality." *Mind* LXIX, no. 273 (1960).

"Modalities and Intensional Languages." *Synthese* XIII, no. 4 (1961).

"Discussion of the Paper of Ruth Barcan Marcus." *Synthese* XIV (1962), by W. V. Quine, S. Kripke, D. Føllesdal, J. McCarthy, and R. B. Marcus.

"Interpreting Quantification." *Inquiry* 5, no. 3 (1962).

"Classes and Attributes in Extended Modal Systems." *Proceedings of a Colloquium on Modal and Many-Valued Logic, Acta Philosophica Fennica* 16 (1963).

"Reply to Dr. Lambert." *Inquiry* 6, no. 4 (1963).

"Iterated Deontic Modalities." *Mind* LXXV, no. 300 (1966).

"Essentialism in Modal Logic." *Noûs* I, no. 1 (1967).

"Modal Logic." In *Contemporary Philosophy,* Vol. 1, ed. R. Klibansky. Florence: La Nuova Italia, 1968.

"Essential Attribution." *Journal of Philosophy* LXVIII, no. 7 (1971).

"Quantification and Ontology." *Noûs* VI, no. 3 (1972).

"Classes, Collections, and Individuals." *American Philosophical Quarterly* (July 1974).

"Does the Principle of Substitutivity Rest on a Mistake?" In *The Logical Enterprise,* ed. R. B. Marcus, A. R. Anderson, and R. M. Martin. New Haven, Conn.: Yale University Press, 1975.

"Wiggins on Identity, Necessity, and Physicalism." In *Philosophy of Logic,* ed. S. Körner. Oxford: Blackwell, 1976.

"Dispensing with Possibilia." Presidential Address. *Proceedings of the American Philosophical Association* 49 (1976).

"Critical Review of L. Linsky *Names and Descriptions.*" *Philosophical Review* (July 1978).

"Nominalism and the Substitutional Quantifier." *The Monist* 61, no. 3 (1978).

"Uninformed Consent." with B. Kucklick and S. Bercovitch. *Science* 205 (1979).

"Moral Dilemmas and Consistency." *Journal of Philosophy* LXXVII, no. 3 (1980).

"Hilpinen's Interpretations of Modal Logic." In *Logic and Philosophy,* ed. G. H. von Wright. The Hague: Nijhoff, 1980.

"Modal Logic, Modal Semantics and Their Applications." *Contemporary Philosophy,* Vol. 1, ed. Fløistad. The Hague: Nijhoff, 1981.

"A Proposed Solution to a Puzzle about Belief." *Midwest Studies in Philosophy: Foundations of Analytic Philosophy,* Vol. VI, ed. P. French et al. (1981).

"Rationality and Believing the Impossible." *Journal of Philosophy* LXXX, no. 6 (June 1983).

"Bar-On on Spinoza's Ontological Proof." *Spinoza: His Thought and Work,* ed. N. Rotenstreich and M. Schneider. Jerusalem: Israel Academy of Science and Humanities, 1983.

"Spinoza and the Ontological Proof." *Human Nature and Knowledge,* ed. A. Donagan, A. N. Perovich, Jr., and M. V. Wedin. Dordrecht: Reidel, 1985. (An expanded version of the preceding paper.)

"Is There Irrationality in the Existence of a Plurality of Philosophical Theories?" *Dialectica* 39, no. 4 (1985).

"On Some Post-1920s Views of Russell on Particularity, Identity, and Individuation." In *Mérites et Limites des Méthodes Logiques en Philosophie,* ed. J. Vuillemin. Paris: Librairie Philosophique, 1986.

"Possibilia and Possible Worlds." *Grazer Philosophische Studien* 25–6 (1985–6).

"A Backward Look at Quine's Animadversions on Modalities." In *Perspectives on Quine,* ed. R. Barrett and R. Gibson. Oxford and Cambridge, Mass.: Blackwell, 1990.

''Some Revisionary Proposals about Belief and Believing.'' *Philosophy and Phenomenological Research* L (Supplement, 1990).

''Ontological Implications of the Barcan Formula.'' In *Handbook of Metaphysics and Ontology.* Munich: Philosophia, forthcoming.

Contributors

Paul Bartha is a Teaching Fellow in Philosophy at the University of Pittsburgh.

Nuel Belnap is Professor of Philosophy at the University of Pittsburgh.

Maxwell J. Cresswell is Professor of Philosophy at Victoria University of Wellington.

Joel Feinberg is Professor of Philosophy at the University of Arizona.

Kit Fine is Professor of Philosophy at the University of California, Los Angeles.

Robert J. Fogelin is Professor of Philosophy at Dartmouth College.

Philippa Foot is Honorary Fellow of Somerville College, Oxford University, and Griffin Professor Emeritus at the University of California, Los Angeles.

Patricia S. Greenspan is Professor of Philosophy at the University of Maryland.

David Kaplan is Professor of Philosophy at the University of California, Los Angeles.

Isaac Levi is Professor of Philosophy at Columbia University.

Charles Parsons is Professor of Philosophy at Harvard University.

Terence Parsons is Professor of Philosophy at the University of California, Irvine.

Scott Soames is Professor of Philosophy at Princeton University.

Ernest Sosa is Professor of Philosophy at Brown University.

Robert Stalnaker is Professor of Philosophy at the Massachusetts Institute of Technology.

David Wiggins is Professor of Philosophy at Birkbeck College of the University of London.

MODALITY, MORALITY, AND BELIEF

I

Modality

1

Ruth Barcan Marcus and the Barcan Formula

TERENCE PARSONS

This smart, well-spoken and beautiful woman enjoyed among her students and also in wider circles wondering admiration, . . . ; she was also sought out by holders of political powers and enjoyed great influence. Morally without blemish, she moved with secure freedom even among men.

said of Hypatia, 4th–5th century logician[1]

Ruth Barcan Marcus is and will remain famous for the Barcan formula. The questions I want to address are these: What is it? Did she really invent it? and Is it worth being famous for? (There is no need for drama; the answers to the last two questions are yes.) I will not discuss the question of whether she deserves to be known for other things as well; the answer to that is also yes, but that is beyond the scope of this essay.

What Is the Barcan Formula?

Part of Ruth Barcan's dissertation project was to give a systematic axiomatic treatment of modal logic as developed by Lewis and Langford in their 1932 textbook, including especially the predicate calculus part containing quantifiers. She did this, and published the results in a relatively new journal, the *Journal of Symbolic Logic* (volume 11), in 1946. This is the first systematic published work that addressed the question of combining unrestricted quantification with unrestricted modality, and the Barcan formula is an essential

I first met Ruth Marcus in September 1965, when I walked into her office to introduce myself. This was not a social call; she was my new boss, having hired me fresh out of Stanford to a brand new department in the newly opened Chicago campus of the University of Illinois. I'll never forget that meeting. She was pleasant, charming, gracious, hospitable, and she scared me to death. What scared me was her asking me whether my dissertation advisor was urging me to publish my thesis. He wasn't. It wasn't good enough. The message was clear: I was in the big leagues. And I wasn't sure I could measure up. Fortunately, it wasn't long before Ruth was letting me know that in her eyes I was measuring up, and that was a great comfort to me. That combination, of setting high standards while also being supportive is just what one wants in a mentor, and I can admit today that I needed a mentor then. I thank Ruth for filling that role for me.

part of her account. The formula is

$$\Diamond(\exists x)S \equiv (\exists x)\Diamond S,$$

or, equivalently, in terms of the universal quantifier and necessity:

$$\Box(x)S \equiv (x)\Box S.$$

Aficionados know that in one direction these biconditionals are called the Barcan formula, and, in the other direction, the converse Barcan formula, but most people (myself included) can rarely remember which is which. Marcus endorsed both directions, and the formula is important and controversial in both directions. (In the first example, the left-to-right implication is the Barcan formula and the converse implication is the converse Barcan formula. Conversely, in the latter formula the right-to-left implication is the Barcan formula and the converse implication is the converse Barcan formula.)

These formulas are important for two reasons. One is the use of notation that clearly distinguishes two different readings, and the other is a proposal for how they are related.

Background

To appreciate the significance of the Barcan formula, we need to know something about the historical background of how modalities related to quantification before 1946. The interaction between modality and quantifiers was discussed by Aristotle, who considered the difference between

An animal is necessarily moral.

and

Necessarily, an animal is mortal.

Aristotle's terminology for the distinction was *composition* and *division*, which is essentially one of grammatical scope. This terminology was used for centuries, and persists today. It is a relatively theory-neutral way to mark the distinction. A new step was taken near the end of the twelfth century when an anonymous writer suggested that the first reading could be considered to attribute a necessity to a *thing*, and the latter to attribute necessity to *what is said*; the distinction came down to us as the contrast between *de re* and *de dicto* modality.[2] The distinction was carefully respected by the great logicians of the thirteenth and fourteenth centuries, though it became less central as logic languished in the centuries thereafter, and you will find what we would call *de re/de dicto* confusions throughout the modern era.

None of this earlier work that respected the *de re/de dicto* distinction was in a position to address the issue Marcus addressed, for two reasons. First, the authors usually had in mind not logical necessity, but something else, something that we would see as more restricted. Second, before Frege, unrestricted quantification was unknown; all logicians used restricted quantification. Let me expand on this latter point. The most sophisticated treatment of quantification prior to Frege was found in supposition theory, as developed by people like Ockham, Buridan, and Paul of Venice. Here is an example of the sort of theoretical apparatus they were dealing with. The term 'donkey', they would say, has *determinate supposition* in the sentence:

Some donkey is brown.

This is because from that sentence one may infer:

This donkey is brown, or that donkey is brown, or that one, and so on for all the donkeys.

This is like saying that 'donkey' is existentially quantified, with wide scope. Similar tests were used for distributive supposition of a term (something like being universally quantified, though not necessarily with wide scope) and merely confused supposition.

It is still not clear today just what was supposed to be accomplished by analyses of the suppositional status of terms, but it is apparent from the sort of tests that were used that an essential precondition for the application of supposition theory is that its goal is a study of restricted quantification. The theory does not classify quantifiers regarding quantity and scope; it classifies terms (predicates). In 'Some donkey is an animal' the occurrence of the quantifier word 'some' is what *gives* the word 'donkey' determinate supposition, but the thing that *has* determinate supposition is the word 'donkey', not the word 'some'. This is a theory that classifies the words that restrict quantifiers, not the quantifiers themselves, and so it is essentially a theory of restricted quantification. It was only when Frege tore apart the quantifiers from their restrictors that unrestricted quantification came to exist, when he symbolized 'Some donkey is an animal' as '$(\exists x)(Dx \,\&\, Ax)$', and then gave a semantics for the quantifier apart from the conjunction. Before Frege, the Barcan formula could not have been written.

Frege, of course, was unknown by most English-speaking philosophers, and it was only after Russell and Whitehead's *Principia Mathematica* that attention to something like modern quantification theory got momentum, impelled in part by Russell's provocative applications of it to philosophical puzzles. These often employed something like his theory of definite and indefinite descriptions, which explicitly uses unrestricted quantifiers to analyze restricted ones. A couple of decades later, modalities were being looked at seriously in modern terms by Lewis and Langford in 1932,[3] and fourteen years after that Ruth Barcan published the first systematic treatise mixing logical modality and unrestricted quantification.

How do I know Marcus was first? My evidence is of three kinds. First, I've asked several knowledgeable people about antecedents, and I've searched some of the likely literature myself. That's the least conclusive part of the evidence. Second, I asked Marcus herself. Surely, given her fame in connection with the Barcan formulas, someone would have pointed out to her antecedents, had there by any. Nobody has. Third – and here is proof beyond Cartesian doubt: Alonzo Church says so. In his first essay on sense and denotation he says, ''The first systematic treatments [of modality in connection with quantifiers] are those of Barcan and Carnap (independent, and nearly simultaneous).'' This narrows it down to Barcan and Carnap, and Barcan published first.[4]

Contemporary Uses of the Barcan Formula

The term 'Barcan formula' has come to have at least three different uses:

1. Any mixing of modalitylike things with quantifiers is often referred to as an instance of the Barcan formula. And this is fair. It wasn't just that modalities hadn't been carefully and systematically mixed with quantifiers before Marcus did it; other intentional operators such as 'knows that' and 'ought' hadn't either, and once a paradigm was in the air, people's interest was directed to important issues. Often in such applications the analogs of the Barcan formulas do not hold, but that is not the important point. By providing an examplar, Marcus focused attention on addressing logically and semantically important issues.

2. In a more restricted context, ''Barcan formula'' sometimes refers to the mixing of any alethic modality with any quantification. Here there are still plenty of counterexamples to the formulas, using restricted kinds of necessity and quantification. For example, consider the inference:

No existing American can run a mile in under 3:45.
∴ It's impossible for an American to run the mile in under 3:45.

If you reject this inference, then you do not accept what is sometimes called the Barcan formula for this case. This is the Barcan formula extended so as to apply to old-fashioned restricted quantification and restricted modality; the quantification is restricted to Americans, and the modality is restricted to what is possible consistent with laws of physiology, gravity, and existing constitutions of human bodies. Construed in such ways, the Barcan formula usually does not hold. But, as I have noted, this is not what Marcus proposed in her original papers. She had in mind unrestricted modality and quantification.

The restricted versions can be expressed in logical notation in ways well known from Frege and Russell, something like this:

(x) (x is an American → □ (physical laws are such and such →
 ¬ (x runs a mile in under 3:45))
∴ □ (physical laws are such and such →
 (x)(x is an American → ¬(x runs a mile in under 3:45))

Nowhere is the pure Barcan formula to be found here. The viability of the Barcan formula as a law of pure logic for unrestricted quantification and modality is untouched by such examples.

3. The third use of the term 'Barcan formula' is its strict use in its original application to unrestricted modality and quantification. And here, it may be completely defensible. This is controversial, because many contemporary philosophers think the Barcan formulas are incorrect as originally proposed. I think that the jury is still out on the issue. To be specific, the arguments that are currently popularly accepted as disproving the Barcan formulas beg questions by explicitly or implicitly using restricted quantification or restricted modality, or by utilizing logical devices that transcend the framework within which the formulas were originally proposed. For the remainder of this paper I will address these points.

But first let me clarify my defense. The Barcan biconditional is sufficiently controversial that I am not comfortable defending it as an a priori principle of logic. It settles too many issues that are debated by reasonable people to be a logical principle. But that leaves open the possibility that every instance of the formula is true. This is what I will concentrate on.

Putative Counterexample to the Converse Barcan Formula

I begin with a putative counterexample to the converse Barcan formula, usually attributed to Kripke:

Necessarily, everything exists. TRUE
∴ Everything necessarily exists. FALSE

In symbols:

□ (x)(x exists) TRUE
∴ (x) □ (x exists) FALSE

The converse Barcan formula endorses this inference, and it allegedly leads from truth to falsehood.

Is this a good counterexample? I think not. It plays on an equivocation in 'exists'. To see this, begin by asking why we think the premise is true. Virtually all nonphilosophers as well as most philosophers do not think that everything exists; numbers don't exist, properties don't exist, and so on. So it is doubtful that many people would agree that necessarily, everything exists. The premise is apparently not plausible at all. The rejoinder to this is that people who object to the premise are taking 'exist' in a restricted sense, one that does not exhaust the range of the quantifiers. To get the counterexample, you can't restrict 'exists' at all. This can be made precise by replacing 'x exists' by '$(\exists\, y)(y = x)$'. The counterexample then becomes

$\Box\,(x)(\exists y)(x = y)$ TRUE
$\therefore\,(x)\,\Box\,(\exists y)(x = y)$ FALSE

So construed, the premise is plausible. But now why reject the conclusion? The conclusion no longer says 'Everything has necessary existence', it says 'Everything is necessarily something' and that sounds plausible, not implausible. After all, everything is necessarily itself, so why isn't everything necessarily something? Remember, being something does not *mean* being something that exists; we are being careful not to sneak in the assumption that if I am necessarily something then I necessarily exist. The Barcan formula only gets us to the claim that I am necessarily something, it does not get us the further inference from that claim to the claim that I necessarily exist. This further inference is accepted by many (though not all) current possible worlds theorists, but it is typically there by assumption, not by argument.

I have argued this way before, so I am aware of the usual rejoinder. This is to note that quantifiers have domains, and that my inference from 'Everything is necessarily itself' to 'Everything is necessarily something (namely, itself)' illegitimately imports a domain along with the 'something'. Quantifiers have domains, these change when you go from world to world, and their presence needs to be taken account of.

Suppose this is right; where does it get us? This requires a slight excursion into domains. Suppose I say, "Everything's calm today," and you reply with "Oh yeah, what about that tenement fire in Minsk?" That is probably not a pertinent rejoinder. You have countered my claim:

$(x)\,(\text{Calm } x)$

with a true claim:

$\neg\,(x)\,(\text{Calm } x)$

but your claim and mine are not inconsistent in spite of appearances; my domain is local events, and your is everything on earth. How can logic handle such phenomena? One way is to go contextual, and to claim that the relativity of domains to quantifiers depends on context. On this approach there is no straightforward application of the logical inferences we teach about in logic courses. I am not at all opposed to this approach, but we really do not have a well-developed theory of how it works. Another equally good approach is to handle such examples by being explicit about the domains. When we say 'everything' we symbolize it as 'every *thing*', and then if you use the same word with a different meaning from my word, we use different words to symbolize our remarks. My claim is symbolized:

$(x)(\text{Thing}_1\ x \rightarrow \text{Calm}\ x)$

and yours as

$\neg\ (x)(\text{Thing}_2\ x \rightarrow \text{Calm}\ x)$

Then your response is not pertinent to mine because you mean something different by 'Thing_2' than I mean by 'Thing_1'.

We can do modal logic in this way as well, being explicit about the domains of our quantifiers. The counterexample proposed before is then symbolized as

$\Box\ (x)(x \text{ is a thing} \rightarrow (\exists y)(y \text{ is a thing} \ \&\ x = y))$ TRUE
$\therefore (x)(x \text{ is a thing} \rightarrow \Box\ (\exists y)y \text{ is a thing} \ \&\ x = y))$ FALSE

The domains then change from possible world to possible world in the same way that the extensions of predicates do.

I now agree that the premise is true and the conclusion false; this argument is not a good one. But it is also no longer a counterexample to the Barcan formula, because it is no longer an instance of the Barcan formula. The introduction of explicit domains has turned the example into one containing restricted quantification. I am not aware of any way to improve on this line of argumentation without begging the question. (For example, moving the modal sign in the conclusion to just inside the initial quantifier produces an example about which partisans will naturally disagree.)

Putative Counterexample to the Barcan Formula

Let me turn now to the argument that most people see as the clincher against the Barcan formula proper. Couldn't there be things different from any actual thing? Not different in kind, but different numerically? If so, don't we have an easy refutation of the Barcan formula? Suppose we use 'actual' to pick out things in the domain of the actual world, and suppose that we alter our modal logic to treat this predicate in a logically special way: It is to be unaffected by modal operators. The counterexample is then:

Possibly, there is something different from any actual thing.
∴ There is something that is possibly different from any actual thing.

Allegedly, the premise is true, because the actual world is not necessarily as large as any world can be; some possible world is bigger, and so it has objects in it that are different from any actual ones. And the conclusion is false because each actual thing is necessarily itself, and so any actual thing is necessarily a thing that in the actual world is actual.

$\Diamond\ (\exists x)\,(x \text{ is a thing } \&\ (y)(y \text{ is actual} \rightarrow x \neq y))$ TRUE
$\therefore\ (\exists x)\ \Diamond\ (x \text{ is a thing } \&\ (y)(y \text{ is actual} \rightarrow x \neq y))$ FALSE

This bypasses any maneuvers with domains. However we read '*x* is a thing', the conclusion follows from the premise by the Barcan formula, and, it is alleged, when '*y*' is restricted in possible worlds to things that are actual in the actual world, the premise is true and the conclusion false.

There are two natural responses to this alleged counterexample. First, it is important to point out the inconclusiveness of the premise. It assumes that this world does not necessarily contain as many things as it could. But is this so? Much recent work in set theory has concentrated on large cardinal assumptions, and there is a natural inclination to demand largeness without limit. Maybe the actual world does contain as many things as a world could possibly contain. And if the existence of pure sets is a matter of necessity, then every world contains as many things as it can. This, together with a large dose of antiessentialism, undercuts the claim that there might have been things in addition to the things that there are.[5]

Suppose you do not find the preceding speculation persuasive. There is still room to question the significance of the counterexample to the Barcan formula. This proposed counterexample is reminiscent of arguments based on Cambridge change. Cambridge change is like this: Suppose that yesterday you were not equal in size to me, but today you are. I didn't change height, but you grew. Then, in a sense, I didn't change, though something is true of me now that was not true of me yesterday. I underwent a Cambridge change: I changed without there being a change in me.

Some people are skeptical of essentialism. Perhaps necessarily cyclists are two-legged (*de dicto*), but is each cyclist two-legged (*de re*) of necessity? Certainly not. Nor are donkeys necessarily animals, if necessity is construed *de re*.[6] This denial of essentialism has wide (though not universal) support. But then people like Plantinga come along and argue that you *have* to endorse essentialism.[7] Here, for example, is an essence of Ruth Marcus: It is true of her in every possible world that she is a philosopher *in the actual world*. It's true in each world that she's a philosopher in this one. Is that essentialism? That's Cambridge essentialism! It's an artificial construction that fits the formula for essentialism: a property that Marcus has in every world. That is the sort of thing that is going on in the counterexample to the Barcan formula. We have a predicate, 'is actual', that is true of a thing in a possible world if something is true of it in this world. We express this Cambridge essence in our language, and as a result, the Barcan formula turns out to be false. Being actual in this sense is a nonlogical essence, but it's a Cambridge essence.

If the only way to conclusively refute the Barcan formula is to expand the notation of modal logic so as to create logically special predicates so as to be able to express Cambridge essences in the object language, then maybe there is something deeply true about what is under attack, something that we have not yet completely fathomed.

NOTES

1. Pauly's *Real-Encyclopädie der Classichen Altertumswissenshcaft*, entry for "Hypatia."
2. Personal communication from Calvin Normore, based on a manuscript translated by Norman Kretzman.
3. C. I. Lewis and C. H. Langford, *Symbolic Logic* (New York: Dover, 1932).
4. The Church reference is to "A Formulation of the Logic of Sense and Denotation," P. Henle, H. M. Kallen, and S. K. Langer (eds.), *Structure, Method and Meaning* (New York: The Liberal Arts Press, 1951); p. 22 , n. 23. Marcus's article is "A Functional Calculus of First Order Based on Strict Implication," *Journal of Symbolic Logic* 11 (March 1946; submitted in September 1945); Carnap's is "Modalities and Quantification," *Journal of Symbolic Logic* 11 (June 1946; submitted in November 1945).
5. Antiessentialism is required as follows. Certainly there might have been more porcupines than there are. So there must be another possible world in which there are things that are porcupines that are not porcupines in this world. The cardinality assumption guarantees that the alternate world has enough things in it to play the roles of the additional porcupines, but the additional things will not be porcupines in the actual world. We thus need the possibility that a thing that is not a porcupine in this world is one in the other world; that thing thus cannot essentially be or not be a porcupine. The point becomes more poignant if we assume that there could be more concrete objects in another world than in this one, thus requiring (by a similar argument) that being concrete or abstract cannot be essential to at least some things.
6. William Ockham even attributes to Aristotle the view that it is false that men are necessarily animals, when this is read *de re* (when it is read as equipollent to a modal proposition in the sense of division). See A. Freddoso and H. Schuurman (trans.), *Ockham's Theory of Propositions* (Notre Dame, Ind.: University of Notre Dame Press, 1980), pp. 111–12.
7. *The Nature of Necessity* (Oxford: Clarendon Press, 1974).

2

The Interaction of Modality with Quantification and Identity

ROBERT STALNAKER

1. Introduction

One might once have hoped that an adequate quantified modal logic could be obtained simply by putting propositional modal logic together with the extensional theory of quantification and identity. One might have expected, on the one hand, that the sentences and schemata valid in the separate theories should remain valid when each was generalized by being combined with the other, and one might have hoped, on the other hand, that any new validities that depended on the interaction of modal and quantificational concepts would be derivable from the axioms and rules of the separate logics for these concepts. But it did not happen that way. The increase in expressive power gained by combining the two languages cast doubt on familiar logical principles such as Leibniz's law and existential generalization, and raised new questions about the logical status of sentences in which modality and quantification were combined. The development of a semantics for the combined languages required decisions in response to questions not raised by either of the separate theories, for example questions about the relationship between the domains of individuals relative to which the quantifiers are interpreted and the domains of possible worlds relative to which the modal operators are interpreted. These questions, and the arguments about how to answer them, helped to clarify the concepts involved in both theories – the nature of variable binding, quantification, and identity, and of the possible worlds used to interpret modality.

Ruth Barcan Marcus was one of the first to explore the issues concerning the interaction of modality and quantification, and the formulas named for her – the Barcan formula and its converse – are the most prominent examples of principles that make general claims that depend essentially on the interaction of modality and quantification. Her discussions of the logical status of these principles, and of other principles in which modal and extensional concepts interact, such as the necessity of identity, helped to clarify the conceptual foundations of quantified modal logic.[1] My aim in this paper is to continue this kind of discussion by looking at two other logical principles that combine modality with quantification and with identity: first, a weaker version of the converse Barcan formula, and second, the principle of the necessity, not of identity, but of distinctness. I will show, first, that while each of these principles is validated by a very natural and very general semantic theory, they are each independent principles – not derivable in a theory that results from combining complete axiomatizations of extensional predicate logic and propositional modal logic. I will argue that there are conceptual assumptions that lie behind and help to explain the independence of these principles, and a semantics with some conceptual interest that invalidates them.

To sharpen the focus on the problems that arise from the interaction of quantifiers with modality, I will begin by sketching and motivating the separate modules – semantics and proof theory first for extensional quantification theory and then for propositional modal logic – before putting them together. My aim is to try to make the generalization that results from combining them as smooth and natural as possible, and to ensure that our logics of predication, quantification, and identity do not include principles that have to be taken back when the expressive power of the language is increased. After sketching semantics and proof theory for the separate theories and putting them together, I will introduce and discuss the qualified converse Barcan formula. While it will be obvious that this principle is valid in the standard semantics, I will show that it is an independent principle by sketching a variant semantics – a version of counterpart theory – that invalidates it while validating the rest of the logic. Finally, I will look at the interaction of modality with identity, showing that the necessity of distinctness is also an independent principle, even in a theory that includes the qualified converse Barcan formula.

2. Extensional Predicate Logic

Our extensional theory is a first-order predicate logic with primitive singular terms and identity. The semantics is standard in most respects, but it has two features that distinguish it from more familiar formulations of predicate logic: First, it has a variable-binding abstraction operator; second, it is a free logic, allowing that the domain may be empty, and that singular terms may fail to denote. The syntax and semantics for the language are spelled out in an appendix. I will here just make a brief comment about each of these idiosyncrasies.

First, on abstraction: The only primitive variable-binding operator in our language is the abstraction operator for defining complex predicates. The quantifiers do not bind variables, but are operators on one-place predicates. This is not the usual way to formulate first-order predicate logic, but it is a minor variation on the usual procedure that gives the same truth conditions to quantified sentences, and that uses exactly the same Tarskian device: Open sentences are interpreted in terms of satisfaction by a sequence rather than truth, and expressions with bound variables are interpreted in terms of satisfaction by certain variant sequences. But while the variable-binding device used in our semantics is essentially the same as the one standardly used, I think the formulation I will use gives a clearer representation of the logic of quantification because it separates two conceptually distinct operations that are performed by variable-binding quantifiers. First is the implicit formation of complex predicates from complex sentences by introducing blanks – free variables – in the sentences. The second is generalization: the formation of general claims from predicates – the claims that everything in the domain, or at least one thing in the domain, satisfies the predicate that is implicitly represented by the open sentence. In our language, the abstraction operator makes the first of these operations explicit, turning an open sentence into an expression that has the syntactic role as well as the semantic function of a predicate. Then the quantifier has only the job of expressing generality.[2]

So rather than a rule for the universal quantifier like this[3]:

If ϕ has the form $\forall x\psi$, then $v^s(\phi) = 1$ iff $v^{s[d/x]}(\psi) = 1$ for all $d \in D$,

we have the following two rules, which together achieve the same effect:

If F has the form $\hat{x}\psi$, then $v^s(F) = \{d \in D: v^{s[d/x]}(\psi) = 1\}$, and
If ϕ has the form $\forall F$, then $v^s(\phi) = 1$ iff $v^s(F) = D$.

Second, a comment on free logic: The models for our language, like those in the classical extensional semantics, are pairs $\langle D, v\rangle$ where D is a set – the domain – and v is a function assigning values to the primitive descriptive expressions – the sentence letters, predicates, and individual constants. The only difference is that the domain D may be empty, and individual constants may lack values. That is, the valuation function v may be a partial function, undefined for some individual constants (although it must be defined for all predicates and sentence letters). To interpret sentences containing nonreferring names, a semantic decision must be made. Our theory makes the simplest decision: Sentences of the form Ft (or more generally of the form $Rt_1 \ldots t_n$) are false if t (or any of $t_1, \ldots, t_n$) is undefined. Some might have preferred a different decision (for example, stipulating that such sentences are neither true nor false, rather than false), but there is nothing here to argue about, so long as one avoids equivocation.[4]

The semantics for identity in our theory is perfectly standard: Identity is the binary relation that is satisfied by every pair $\langle d, d\rangle$ such that d is a member of the domain. But since names may fail to refer, and since predication sentences with nonreferring terms are stipulated to be false, some identity theses normally assumed to be valid will be invalid, for example, $t = t$. But the universal generalization of this, $\forall\hat{x}(x = x)$, will of course be valid, since the property of being self-identical must be true of everything.

Both of the idiosyncratic features of our extensional predicate logic are included in anticipation of the modal generalization in which the problems they address require attention. The abstraction operator provides the resources to express modal properties and to make scope distinctions that involve modality and predication, but not quantification. The elimination of existential presuppositions makes room for the possibility that individuals not exist necessarily by providing for the interpretation of sentences containing singular terms that have no referent in the domain of the possible world in which they are being interpreted. But even though the theory has features that are motivated by the modal generalization, the language, logic, and semantics remain purely extensional.

Here is a complete axiomatization of this version of free predicate logic:

Axiom schemata:

Propositional Logic:	All tautologous sentences are axioms
Abstraction:	$\vdash \forall\hat{x}(\hat{y}\phi x \leftrightarrow \phi^{x}/_{y})$[5]
Quantification:	$\vdash \forall\hat{x}(\phi \rightarrow \psi) \rightarrow (\forall\hat{x}\phi \rightarrow \forall\hat{x}\psi)$
	$\vdash \forall\hat{x}(\forall F \rightarrow Fx)$
	$\vdash \forall\hat{x}Fx \rightarrow \forall F$ (provided x is not free in F)[6]
	$\vdash \forall\hat{x}\forall\hat{y}\phi \rightarrow \forall\hat{y}\forall\hat{x}\phi$[7]
Existence:	$\vdash F^{n}t_1 \ldots t_n \rightarrow \exists\hat{x}x = t_1$
Identity:	$\vdash s = t_i \rightarrow (F^{n}t_1 \ldots t_i \ldots t_n \rightarrow F^{n}t_1 \ldots s \ldots t_n)$[8]

Rules:

Modus Ponens:	If $\vdash \phi \rightarrow \psi$ and $\vdash \phi$, then $\vdash \psi$
Universal Generalization:	If $\vdash \phi \rightarrow \psi$, and there are no free occurrences of x in ϕ, then $\vdash \phi \rightarrow \forall\hat{x}\psi$

Most of these axioms and rules will be familiar (at least if one ignores the "^" in expressions of the form $\forall\hat{x}\phi$). "ϕ" stands for any sentence, open or closed. "F" for any one-place predicate, "F^n" for any n place predicate, "s" and "t_i" for any singular terms (individual constants or variables), "x" and "y" for any individual variables. Predicates may be simple or complex, open or closed, but one cannot treat sentences generally as predications. For example, "$\hat{x}(Hx \wedge Gx)s$" is an instance of the form Fs, but "$(Hs \wedge Gs)$" is not. The identity axiom schema is thus much more restricted than Leibniz's law as usually formulated. All instances in the extensional language of the general schema, $s = t \rightarrow (\phi \rightarrow \phi^{s}/_{t})$, will be theorems of the extensional logic, but they are not instances of the identity-axiom schema. The validity of the general schema, unlike the validity of the identity axioms, depends on the expressive limitations of the extensional theory, and when the language is enriched, some new instances of it will not be theorems.

3. Propositional Modal Logic

The propositional modal language is a perfectly standard one with sentence letters, truth-functional connectives, and one modal operator "$\Box$" for necessity. The semantics is the standard Kripke semantics: A model consists of a frame or model structure, $\langle W, R \rangle$, where W is a nonempty set of possible worlds and R is a binary relation (the accessibility relation) on them – and a valuation function v that assigns truth values (1 or 0) to the sentence letters at each of the possible worlds. This modal semantics is a generalization of extensional semantics, doing for many possible worlds at once what the extensional semantics does for one world – the actual world. Since the extensional semantics assigns truth values to sentences, the modal generalization assigns truth values to sentences at each possible world. The same rules for the truth functional connectives give truth values for complex sentences at each possible world as a function of the truth values at the worlds of their parts. Necessity is truth in all accessible possible worlds. That is, $\Box \phi$ is true in w iff ϕ is true in all possible worlds accessible to w. ($v_w(\Box \phi) = 1$ iff $v_u(\phi) = 1$ for all u such that wRu) In the most general version of this semantics, no constraints are put on the accessibility relation relative to which the necessity operator is interpreted. The logic validated by this minimal semantics, K is as follows:

Axiom Schemata:

$\vdash$ All tautologous sentences are axioms
$\vdash \Box(\phi \rightarrow \psi) \rightarrow (\Box \phi \rightarrow \Box \psi)$

Rules:

Modus Ponens: If $\vdash \phi \rightarrow \psi$ and $\vdash \phi$, then $\vdash \psi$
Necessitation: If $\vdash \phi$, then $\vdash \Box \phi$

4. Combining the Theories

The basic idea of the modal semantics provides a way to generalize any extensional theory to a modal version of that theory. Whatever the extensional semantics does to define the extensions of complex expressions as a function of the extensions of their parts, just do the same for many possible worlds at once. Whatever an *extension* is according to the relevant extensional theory, let an *intension* in the modal generalization of that theory be a function from possible worlds into extensions. So to combine our two separate theories, we need models that provide a set of possible worlds and for each world, whatever the extensional semantics provided. In our extensional semantics for predicate logic, a model is a domain of individuals and a valuation function that assigns extensions to the primitive descriptive expressions: truth values, members of the domain, and sets of *n*-tuples of members of the domain to the sentence letters, individual constants, and *n*-place predicates, respectively. So the models in the modal semantics should assign to each possible world a domain, and to the sentence letters, individual constants, and *n*-place predicates, functions from possible worlds to, respectively, truth values, members of the domain of the world, and sets of *n*-tuples of members of that domain.

That is, a model for modal predicate logic is a quadruple, $\langle W, R, D, v\rangle$. W and R are the same as in the semantics for propositional modal logic; D is the domain function from W into sets of individuals. For $w \in \mathrm{W}$, D_w is the domain of possible world w. The valuation function assigns *intensions* to the descriptive expressions, where an intension is a function from possible worlds into extensions, and extensions are as defined in the extensional semantics. The semantic rule for the necessity operator remains exactly the same, and the rules for the logical expressions of the predicate logic are straightforward generalizations of the extensional rules: Just add a subscript "w" to the v's and D's in each rule. For example, the rule for universal generalizations is as follows:

If ϕ has the form $\forall F$, then $v^s_w(\phi) = 1$ if $v^s_w(F) = D_w$, and 0 otherwise.

The only place where the generalization from extensions to intensions is not perfectly straightforward is with the interpretation of open sentences in terms of an assignment function.[9] In the extensional theory, the assignment function assigns values – individuals – to the variables. In the modal generalization, it should still be individuals (and not, say, functions from possible worlds into individuals) that are the values of variables, because open sentences are devices for the formation of complex predicates that express properties of individuals, and not properties of some kind of intension. So in the semantics for the modal predicate logic, the assignment function is a function from variables into possible individuals, where a possible individual is a member of the domain of some possible world. The semantic rule for the abstraction operator is then a straightforward generalization of the extensional rule, obtained as with the other rules simply by adding a subscript for the world to the v and the D:

If F has the form $\hat{x}\phi$, then $v^s_w(F) = \{d \in D_w\!: v^{s[d/x]}_w(\phi) = 1\}$.

The axiomatizations for the separate theories remain sound relative to this generalized semantics: We don't have to take anything back. But we do not get completeness for the proof theory that is the result of simply combining the rules and axiom schemata of the separate theories. It is interesting, I think, to consider why not.

5. A Variation on the Barcan Formula

The Barcan formula (BF) and its converse (CBF) are principles that express claims that involve the interaction of the universal quantifier with the necessity operator.

$$\text{(BF) } \forall\hat{x}\Box\phi \rightarrow \Box\forall\hat{x}\phi \quad \text{and} \quad \text{(CBF) } \Box\forall\hat{x}\phi \rightarrow \forall\hat{x}\Box\phi$$

It was for a long time unclear whether these principles were provable from unassailable logical principles, and whether they ought to be theorems of an adequate quantified modal logic. But Kripke's semantics made clear exactly what semantic assumptions were required to validate the principles, and he diagnosed the fallacy in proofs that purported to deduce them from premises that did not make the required assumptions.[10] Both principles are valid only if one makes assumptions about the relations between the accessibility relation and the domains. Specifically, BF is valid if it is assumed that whenever wRu, $D_u \subseteq D_w$, and CBF is valid if it is assumed that whenever wRu, $D_w \subseteq D_u$. Neither assumption seems intuitively motivated in general, although either or both might be plausible in some particular application, or with some particular interpretation of the accessibility relation. But QCBF, a qualified version of the converse Barcan formula does seem to be validated without any assumptions about the relationships between the domains of the different possible worlds:

$$\text{(QCBF)} \quad \Box\forall\hat{x}\,\phi \rightarrow \forall\hat{x}\Box\ (Ex \rightarrow \phi)$$

where "E" is the predicate of existence (defined as $\exists\hat{y}(x = y)$). Whatever the relations between the domains, surely if in w it is necessary that everything satisfies ϕ, then anything that exists in w must satisfy ϕ in every accessible possible world *in which that individual exists.* It is a straightforward matter to show that QCBF is valid in the semantics sketched, but it is not a theorem. To show that it is not, I will construct a variant semantics that validates the axioms and rules of the theory, but in which QCBF is invalid. After sketching the variant semantics and the counterexample in it to QCBF, I will consider whether the nonstandard semantics might have some conceptual interest beyond its use as a technical device to prove independence.

The variant semantics makes a different decision at the place where the modal generalization was least natural: the interpretation of free variables and open sentences. The central idea of the modal generalization (that modal semantics does for many worlds at once whatever the extensional semantics does for one world) did not give clear directions about how to handle the interpretation of free variables in the general theory, and it put no constraints on the relationships between the individuals in the different domains. Suppose we begin by assuming that the domains are disjoint, where this is motivated not by the substantive assumption that individuals could not possibly exist while being different in any way, but instead by the assumption that the entities in the domains are not individuals themselves, but world-slices, analogous to time-slices, individuals-at-a-world. Even if Socrates might have been hooknosed, and so exists and is hooknosed in some counterfactual world c, it remains true (by definition) that the entity. Socrates-as-he-is-in-this-world manifests itself only in this world. Set aside for the moment your worries about whether there really are any such things as Socrates-as-he-is-in-@, and assume that it is entities of this kind that, according to our variant semantics, populate the domains of the possible worlds. Now when we ask whether Socrates (or Socrates-as-he-is-in-@) has, in @ (the actual world), the modal property being possibly hooknosed, we still want the answer to be yes. Even though (according to the story I am trying to tell) it isn't Socrates-in-@ who is the hooknosed member of the domain of the counterfactual world c, it is still Socrates-in-@ who has, in @, the modal property of being possibly hooknosed. "He" has this property in virtue of his relation to the individual – his counterpart, Socrates-in-c – who, in c, is hooknosed.[11]

Our variant semantics, then, assumes that the domains are disjoint, and that the modal properties of the world-bound individuals that populate the domains are determined by a counterpart function that takes a possible world into the counterpart (if any) of the individual. Our version of counterpart theory is conservative (as counterpart theories go) in the following sense: It is assumed that an individual can have at most one counterpart in any possible world.[12] But the counterpart relation is not just a mirror of the identity relation. We do not assume that it is symmetric or transitive. My counterpart's counterparts need not be mine, and I need not be my counterpart's counterpart.

More explicitly, we add to the semantics, along with the requirement that the domains be disjoint, a partial function f that takes possible individuals and worlds into possible individuals. The only restrictions are that if $d \in D_w$, then $f(d, w) = d$ (every individual is its own counterpart in its own world), and that if $f(d, w)$ is defined, then $f(d, w) \in D_w$. (Any counterpart in a world must be in the domain of that world.)

Surprisingly, this variation requires only one small and localized change in the semantic rules, a change in the rule assigning values to the variables. The original rule was this:

If x is a variable, $v^s_w(x) = s(x)$.

The corresponding rule in the variant semantics is this:

If x is a variable, $v^s_w(x) = f(s(x),w)$.

The assignment function, as before, assigns possible individuals to the variables, but the extension of a variable at a world is the counterpart of the individual that is the value of the variable. With all of the other rules, including the rule for complex predicates, remaining the same, this gets us the result we want: Socrates will have the property of being possibly hooknosed if he has a counterpart who is hooknosed in some accessible possible world.

Because the change in the semantics is so small, and because it does not affect either the semantic rule for the modal operator or any of the rules that concern the relationships between extensions within a possible world, it is easy to check that all of the axioms and rules of inference of the combined modal predicate logic are valid in the variant semantics. All that remains to be done to show the independence of QCBF is to give a countermodel, in the variant semantics, for an instance of that schema.

The instance of the schema to be falsified is this:

$$\Box \forall \hat{x}(Fx \vee \Diamond Gx) \rightarrow \forall \hat{x} \Box (Ex \rightarrow (Fx \vee \Diamond Gx))$$

Here is the model: There are two possible worlds, α and β, which are accessible to themselves and each other. $D_\alpha = \{1,2\}$ and $D_\beta = \{3\}$. $f(1,\beta) = 3$ and $f(3,\alpha) = 2$. (That is, 3 is 1's counterpart in β, and 2 is 3's counterpart in α) $v_\alpha(F) = \{1\}$, and $v_\alpha(G) = \{2\}$. $v_\beta(F) = v_\beta(G) = \emptyset$. In α, since 1 satisfies Fx and 2 satisfies Gx, and so $\Diamond Gx$, it follows that $\forall \hat{x}(Fx \vee \Diamond Gx)$ is true in α. In β, the only individual, 3, has a counterpart to α that satisfies Gx, so it satisfies $\Diamond Gx$; consequently, in β also, $\forall \hat{x}(Fx \vee \Diamond Gx)$ is true. Since $\forall \hat{x}(Fx \vee \Diamond Gx)$ is true in both possible worlds, the antecedent of the sentence to be falsified, $\Box \forall \hat{x}(Fx \vee \Diamond Gx)$ is true in α. But the consequent, $\forall \hat{x} \Box (Ex \rightarrow (Fx \vee \Diamond Gx)$, is false in α, since 1 does not satisfy the open sentence $\Box (Ex \rightarrow (Fx \vee \Diamond Gx))$. Ex is true of 1 in β, since it has a counterpart in β, but Fx is false of 1's counterpart in β, as is $\Diamond Gx$, since 1 has no counterpart that satisfies Gx.

This counterpart semantics makes its technical point – showing the independence of QCBF – whether or not one thinks it has any intuitive application. But I want to suggest that it also makes a conceptual point about the status of this principle. The variant semantics brings to the surface an assumption about the relation between the modal properties of an individual in different possible worlds – an assumption implicit in the standard semantics that is not grounded in the nonmodal logic of predication, or in the modal logic of propositions, or in their combination. The counterpart semantics can be seen as a semantics that is neutral on a question that is answered by the standard semantics, since the standard semantics can be seen as a special case of the counterpart semantics, one that is obtained by putting constraints on the counterpart relation. Suppose we impose two additional conditions on the counterpart function:

(1) $f(f(d,w),w') = f(d,w')$
(2) If $d \in D_w$ and $f(d,w') = d'$, then $f(d',w) = d$.

With these conditions, it will follow that the relation that holds between an individual and its counterparts is an equivalence relation. Individuals in the standard semantics can be identified with equivalence classes of the world-bound individuals in the counterpart semantics, and world-bound individuals in the counterpart semantics can be identified with individual–world pairs, in the standard semantics. Since there is a version of the counterpart semantics that is equivalent to the standard semantics, the philosopher who makes the standard assumptions about individuals can have no complaint with the semanticist who wishes to formulate the theory in a way that is neutral on a question that is not settled by either the modal logic of propositions or by the logic of predication and quantification – a question about the relationship between the modal properties that an individual has in different possible worlds.[13] And he cannot legitimately claim that the counterpart theorist is adopting a nonstandard account of predication or quantification, since he and the counterpart theorist share the same logic and semantics of predication and quantification. The nonstandard conceptions of individuals and their modal properties that the counterpart semantics allows for may not be defensible, but the issue is a philosophical one that cannot be settled by logical theory.

6. Identity and Distinctness

The standard theory and the counterpart theory do not disagree about the semantics of identity: Identity is the relation that holds between any thing and itself, and not between any thing and anything else. And they agree that identity is an essential relation: For any things x and y that are identical, x and y are essentially identical – identical in all possible worlds in which that thing exists (or has a counterpart).[14] How could it be otherwise? If x is y, then there is only one thing. Not only can't it be distinct from itself, it cannot have a counterpart in any world that is distinct from itself.[15]

As one would expect for a principle that is valid in the neutral counterpart semantics as well as the standard semantics, the principle of the essentiality of identity (EI):

$$\text{(EI)} \quad \forall\hat{x}\forall\hat{y}(x = y \rightarrow \Box\,(Ex \rightarrow x = y))$$

is provable, using only the identity axiom, the minimal propositional modal logic, and the extensional logic of predication and quantification. But a stronger principle, the *necessary* essentiality of identity (NEI):

$$\text{(NEI)} \quad \forall \hat{x} \forall \hat{y} \Box (x = y \rightarrow \Box (Ex \rightarrow x = y))$$

is not provable in the minimal logic, and not valid in the counterpart semantics. We can make sense, in the counterpart semantics, of a counterexample to this principle, since we can make sense of distinct individuals that might be identical (by having the same counterpart in some world). Suppose worlds α and β are each possible with respect to each other, and that two distinct things in world α, 1 and 2, have the same counterpart, 3, in world β. Then the pair $\langle 1, 2 \rangle$ satisfies the identity relation in world β, but since 1 exists in world α and is distinct from 2, the pair fails to satisfy $\Box (Ex \rightarrow x = y)$ in β. So there can be two (distinct) things that might be identical without being essentially identical, which falsifies NEI.

Necessary essentiality of identity is of course valid in the standard semantics, and it can be proved with the help of QCBF.[16] But a related principle that is valid in the standard semantics (and not in the counterpart semantics), the necessity of distinctness (ND):

$$\text{(ND)} \quad \forall \hat{x} \forall \hat{y} (x \neq y \rightarrow \Box\, x \neq y)$$

cannot be proved even with the help of QCBF. To show the independence of this principle, I will use a more local semantic variation: a standard interpretation of individuals, but a nonstandard interpretation of the identity relation.

It has been noted, by Quine among others, that identity is indistinguishable from indiscernibility.[17] If we interpret the identity sign, not as identity, but as indiscernibility,[18] then even though distinct things might now stand in the relation expressed by the identity sign, the extensional logic of this relation remains the same. That is, the identity axiom,

$$\vdash s = t_i \rightarrow (F^n t_1 \ldots t_i \ldots t_n \rightarrow F^n t_1 \ldots s \ldots t_n),$$

remains valid under this reinterpretation of the meaning of "=". None of the other axiom schemata or rules (with one exception) include the identity sign essentially, so they are unaffected by a change in the interpretation of this predicate. The exception is the existence axiom,

$$\vdash F^n t_1 \ldots t_n \rightarrow \exists \hat{x} x = t_i,$$

which obviously remains valid when the sign is given the weaker interpretation. So all of our axioms and rules, of both the extensional predicate logic and the propositional modal logic are validated, as is QCBF.

Since the essentiality of identity is a theorem, it must be that the essentiality of indiscernibility is as sound a principle as the essentiality of identity; even the necessary essentiality of indiscernibility must be sound, since NEI can be proved with the help of QCBF. These principles can be seen to be valid: Anything that is possibly discernible (or even possibly possibly discernible) is actually discernible, since possibly discernible things can be distinguished by their modal properties. (If *a* is possibly discernible from *b*, then *a*, but not *b*, *actually* has the property of being possibly discernible from *b*.) But the necessity of *discernibility*, unlike the necessity of distinctness, is not a valid principle. Discernible things are not necessarily discernible, since distinctions present in the actual world may be absent in some accessible possible world.[19] Suppose there are two worlds, α and β, β accessible to α, but α not accessible to β. The fact that two individuals were discernible in α would not ensure that they were discernible in β, since what is true in α cannot affect what is true in β.

Since (at least in the theory that makes the standard assumptions about the modal properties of individuals), the necessity of distinctness is valid when " = " is interpreted as identity, but invalid when it is interpreted as indiscernibility, the combination of identity theory with modality provides the resources to distinguish identity from a weaker relation that cannot be distinguished from it in a nonmodal context. Perhaps this shows that there is in some sense something modal about the concept of identity. This conclusion would not be entirely surprising, since intuitively one is inclined to distinguish identity from indiscernibility in modal terms. Indiscernibility and discernibility, unlike identity and distinctness, are fragile relations. Distinct but indiscernible things are only accidentally indiscernible, owing this relation to the limitations of our powers, or the power of our language, to discern; and discernibility, unlike distinctness, could be lost if we lost an ability to discern, or if distinguishable things changed to become indistinguishable.

7. Summary and Conclusion

If we are careful in the way we formulate the extensional logic of predication, quantification and identity, it turns out that we can come close to the original goal of doing quantified modal logic simply by putting propositional modal logic together with extensional predicate logic and identity theory. We do not have to take anything back when we put the two theories together, but if we do the combined semantics in the usual way, we do have to add two principles that concern the interaction of modality with predication, and modality with identity. These are logical principles that are valid but that cannot be derived from the logical principles governing the separate notions. If we add just these two principles[20] we get a quantified modal logic that is complete relative to the standard semantics. The assumptions implicit in the standard semantics that account for the validity of these two principles can be identified, and we can give a semantics that does not make those assumptions and that is at least formally coherent. In this more neutral semantics for modal predicate logic with identity, the valid sentences are exactly those derivable using the rules and axiom schemata of the separate logics.[21]

Appendix

This is a specification of the language and semantics for the extension predicate logic.

Syntax

Vocabulary. The vocabulary consists of three kinds of primitive descriptive expressions: sentence letters, n-place predicates for each $n > 0$, and individual constants (a denumerable set of each), individual variables (a denumerable set), four syncategorematic logical symbols, "$\forall$", "$\wedge$", "$\sim$", and "$\hat{}$", a two-place logical predicate, "$=$", and parentheses for punctuation. Nonprimitive logical symbols, "$\exists$", "$\vee$" "$\rightarrow$", "$\leftrightarrow$", are to be understood as abbreviations in the usual way. A *singular term* is either a variable or an individual constant.

Formation Rules. There are two kinds of complex expressions, sentences and one-place predicates, defined by the following rules:

If F is an n-place predicate and $t_1 \ldots t_n$ are singular terms, then $Ft_i \ldots t_n$ is a sentence.
If ϕ is a sentence, then $\sim\phi$ is a sentence.
If ϕ and ψ are sentences, then $(\phi \wedge \psi)$ is a sentence.
If F is a one-place predicate, then $\forall F$ is a sentence.
If ϕ is a sentence, then $\hat{x}\phi$ is a one-place predicate.

Semantics

Models and Assignment Functions. A model is a pair $\langle D, v\rangle$, where D is a set, the domain, and v is the valuation function assigning values – extensions – to the primitive descriptive expressions of the language. The domain may be empty, and the valuation function may be a partial function, assigning no values to some singular terms (though it must be fully defined for all sentence letters and predicates).

The extension of a sentence letter is a truth value: 1 or 0; the extension of an n-place predicate is a set of n-tuples of members of D; the extension of an individual constant, if it has one, is a member of D.

Well-formed expressions in general, including both open and closed sentences and predicates – receive values in a model relative to an assignment of values to the variables. The closed sentences and predicates receive the same value for all assignments of values to the variables. An assignment function, *s*, is a partial function from the set of individual variables into the domain. (If the domain is empty, the only assignment function will be the one that is undefined for all values.)

If *s* is an assignment function, then $s[d/x]$ is the assignment function that is everywhere like *s*, except that for *x* it takes the value *d*

The syncategorematic logical symbols are interpreted by the semantic rules, but there is one logical symbol that is a predicate and gets assigned an extension: the identity sign, " = " is a two-place predicate whose value, for all valuation functions, is given as follows: $v(=) = \{ \langle d, d\!:\rangle\ d \in D\}$. Since identity is a two-place predicate, uniform formation rules for atomic sentences imply that identity sentences have the form "$= t_1t_2$" but, bowing to tradition, we will abbreviate them in the usual way: "$t_1 = t_2$".

Semantic Rules. The semantic rules extend the valuation function to give values, relative to an assignment function, to the complex expressions – one-place predicates and sentences. If *v* is the valuation function and *s* the assignment function, then v^s is the function that assigns values relative to the assignment function. If *x* is a variable, $v^s(x) = s(x)$. If ϕ is a primitive descriptive expression, then $v^s(\phi) = v(\phi)$.
The rules are as follows:

Sentences.

If ϕ has the form $F^n t_1 \ldots t_n$, then $v^s(\phi) = 1$ if
$\langle v^s(t_1), \ldots v^s(t_n)\rangle \in v^s(F^n)$. $v^s(\phi) = 0$ otherwise.
If ϕ has the form $\sim\psi$, then $v^s(\phi) = 1 - v^s(\psi)$.
If ϕ has the form $(\psi_1 \wedge \psi_2)$, then $v^s(\phi) = v^s(\psi_1) \times v^s(\psi_2)$.
If ϕ has the form $\forall F$, then $v^s(\phi) = 1$ if $v^s(F) = D$. $v^s(\phi) = 0$ otherwise.

Predicates.

If *F* has the form $\hat{x}\phi$, then $v^s(F) = \{\mathrm{d} \in \mathrm{D}\colon v^{s[d/x]}(\phi) = 1\}$.

NOTES

1. See Ruth C. Barcan, "A Functional Calculus of First Order Based on Strict Implication," *Journal of Symbolic Logic* 11 (1946): 1–46, where a version of the Barcan formula is introduced, and "Modalities and Intensional Language," *Boston Studies in the Philosophy of Science*, vol. 1 (New York: Humanities Press, 1963) for a discussion of the necessity of identity.

2. In "Complex Predicates," *Monist* 60 (1977), I argued in more detail for the conceptual advantages of this way of formulating extensional predicate logic.
3. s is a function assigning values (members of the domain) to the variables, or equivalently, a sequence of members of the domain. "$s[x/d]$" denotes the function that is everywhere like s, except that the variable x is assigned the value d.
4. The truth-value gap decision would require additional decisions about the interpretation of complex sentences containing sentential components that lack truth value. This is the reason that our decision is the simpler, and more conservative one.

   A second alternative decision, sometimes taken in semantics for free logic, would require adding resources to the semantic structure relative to which the sentences are interpreted. This is the Meinongian decision: to add an outer domain of nonexistent individuals, or perhaps a single arbitrary virtual individual that is outside of the domain, to be the referents or referent of singular terms that do nor refer to any existent thing. The nonexistent individuals or individual might or might not be in the extension of any predicate.
5. "$\phi^x/_y$" denotes the result of substituting x for all free occurrences of y in ϕ (relettering bound variables in ϕ if necessary to ensure that all new occurrences of x in $\phi^x/_y$ are free). "x" and "y" are to be understood as schematic metavariables ranging over the variables of the object language.
6. This axiom, necessary to tie up a small loose end, is nonredundant only for the case where F is a primitive predicate.
7. This permutation principle, surprisingly, is not a consequence of the other axioms. The axiomatization for a pure quantified modal logic in Saul Kripke, "Semantical Considerations on Modal Logic" *Acta Philosophical Fennica* 16 (1963) needs to be supplemented with this axiom schema to be complete. See Kit Fine, "The Permutation Principle in Quantificational Logic," *Journal of Philosophical Logic* 12 (1983).
8. Identity theory usually needs an additional axiom – the reflexivity of identity, $\forall \hat{x}(x = x)$. But because of the existence axiom, and the way existence is defined in terms of identity, this principle is a theorem.
9. One might think of the situation this way: The Tarskian semantics for quantification generalizes truth simpliciter to satisfaction – truth of a sequence of individuals, or truth relative to an assignment. Kripkean modal semantics generalizes truth simpliciter to truth at a world. The problem of semantics for quantified modal logic is to fit the two generalizations together to get truth relative to an assignment at a world (or truth at a world, relative to an assignment.)
10. See the paper by Kripke cited in note 7.
11. Counterpart theory was invented by David Lewis, though his version of the theory differs both in intuitive interpretation and in technical detail from the one I am using here. The motivation for Lewis's theory is connected with his modal realism; the world-bound individuals that populate the domains of the possible worlds are ordinary individuals. And Lewis does not assume, as I do, that an individual may have at most one counterpart in any possible world. See Lewis, "Counterpart Theory and Quantified Modal Logic," *Journal of Philosophy* 65 (1968), and for a more philosophical discussion in the context of a defense of modal realism, chapter 4 of Lewis, *On the Plurality of Worlds* (Oxford: Blackwell's, 1986). See also Allen Hazen, "Counterpart Theoretic Semantics for Modal Logic," *Journal of Philosophy* 76 (1979).

    For a philosophical discussion of the actualist interpretation of counterpart theory, see my Counterparts and Identity, *Midwest Studies in Philosophy*, vol. XI: *Studies in Essentialism* (Minneapolis: University of Minnesota Press, 1986).

12. The theory can accommodate a counterpart relation that allows multiple counterparts in a single world by representing such a relation by a set of counterpart functions and letting the semantics be indeterminate between them. If, for example, London has two counterparts, one pretty and one not, in a certain possible world, then if "*F*" expresses the property of being pretty, it will be semantically indeterminate whether Fx is true or false in that world, relative to the assignment of London to x. In general, sentences will be determinately true (false) if and only if they are true (false) for all admissible ways of arbitrarily resolving the indeterminacy. cf. pp. 136–7 of my Counterparts and identity, cited in note 11.
13. One way to put the question in dispute, in terms of an example, is this: Suppose it is agreed that in the actual world, Socrates has the property of being possibly hooknosed, which is to say that there is a possible world in which he is hooknosed. Suppose we agree further that in the world in which Socrates is hooknosed, he has the modal property of being possibly flatfooted. Does it follow that Socrates has, in the actual world, the modal property of being possibly flatfooted – or at least of being possibly possibly flatfooted? The counterpart theorist says that this does not follow without further assumptions that are not grounded in the standard interpretation of "possibly," and the standard assumptions about the way predicates are formed from open sentences.
14. The principle of the necessity of identity, $\forall \hat{x}(x = y \rightarrow \Box (x = y))$, fails in both the standard semantics and the counterpart semantics, but only because a thing may exist contingently, and because self-identity, like all properties, entails existence.
15. A thing might, according to the more radical counterpart theorist, have more than one counterpart in some possible world, and so might have counterparts that are distinct from each other. But if (in the actual world) $x = y$, then there are not two thing in the actual world, the x and the y, one with one counterpart and the other with the other. Even in the more radical theory, there will be no pair of thing of which $x = y$ is true, but $\Box (Ex \rightarrow x = y)$ false.
16. Start with the necessitation of EI, use instances of QCBF to move the necessity sign inside the quantifiers, and then eliminate the Ex and Ey antecedents, using the existence axiom. (Since $\vdash x = y \rightarrow Ex \wedge Ey$, by the existence axiom, $x = y \rightarrow \phi$ follows from $(Ex \rightarrow (Ey \rightarrow (x = y \rightarrow \phi)))$.)
17. See, for example, W. V. Quine, *Word and Object* (Cambridge, Mass.: MIT Press, 1960), p. 230.
18. More precisely, say that two individuals a and b, are indiscernible in a possible world w in a model $\langle W,R,D,v\rangle$ if and only if, for all assignment functions s and all one-place predicates F that have no occurrences of " = " (where F may be simple or complex, open or closed), $a \in v^s_w(F)$ iff $b \in v^s_w(F)$. Then we replace the rule giving the meaning of the identity predicate with the following:

$$v_w(=) = \{\langle a,b\rangle \in D_w\text{: } a \text{ is indiscernible from } b \text{ in } w\}.$$

19. A counterexample to the necessity of discernibility will require a model with a nonsymmetric accessibility relation, which is what allows distinctions to be lost when one moves to an accessible world. In a semantics for S5, or any semantics that assumes the symmetry of accessibility, the necessity of discernibility will be valid, and the necessity of distinctness will be provable in the appropriate modal logic.

20. The schemata we need are generalized versions of QCBF and ND:

$$\vdash \Box^{n} \forall\hat{x}\phi \rightarrow \forall\hat{x} \Box^{n} (Ex \rightarrow \phi)$$

$$\vdash \forall\hat{x} \Box^{m} \forall\hat{y} (x \neq y \rightarrow \Box^{n} x \neq y)$$

where "$\Box^{n}$" stands for n occurrences of "$\Box$".

21. Our propositional modal logic is K, and the accessibility relation is unconstrained. If our logic were instead M, K4, S4, S5, KB or B, and the accessibility relation were constrained in the appropriate way, the completeness claims would still hold. But there are complete propositional modal logics whose predicate logic extensions are incomplete.

3

S1 Is Not So Simple

MAXWELL J. CRESSWELL

Semantical study of what are called nonnormal modal logics, systems that include Lewis's S1–S3 and S6–S8, and their counterparts $S1^0$–$S3^0$ and $S6^0$–$S8^0$, which lack the axiom of necessity $Lp \dashv p$, proceed by distinguishing between 'normal' worlds – where L is evaluated as in 'normal' modal logics like K, T, and the like, and 'nonnormal' worlds, where it is evaluated differently (see chapter 15 of *An Introduction to Modal Logic,* Hughes and Cresswell 1968, henceforth referred to as IML). In the stronger nonnormal systems, S2, $S2^0$, and beyond, the rule is simple. $L\alpha$ is false in a nonnormal world for every wff α. But for S1 thing are not so simple. In Cresswell 1972 I provided a semantics for S1 that treated $L\alpha$ in a nonnormal world by what has come to be called a 'neighborhood' relation (see Segerberg 1971, p. 13). This is a relation between a world and a set of worlds. The sets of worlds related to a given world are its neighborhoods. The completeness proof in Cresswell 1972 used algebraic techniques and the first part of the present paper provides an alternative proof using canonical models.

Since 1972 there have appeared a number of discussions of systems in the vicinity of S1. These discussions have been concerned to develop a relational semantics for such systems, where this means a semantics involving a relation

I cannot pretend that this essay is *directly* connected with Ruth Marcus's work. Yet it is I believe closer to it than most recent work in propositional modal logic. Nowadays, study of modal systems is usually restricted to those which are called 'normal'. Of these the only *Lewis* systems are S4 and S5, which appear in Lewis and Langford 1932 only in an appendix on p. 500f., and are rejected by Lewis as acceptable systems of strict implication. S3 is the system of Lewis 1918, and S1 and S2 are the systems developed in chapter VI of Lewis and Langford 1932. S3 was subsequently discovered to be stronger than S2, but Lewis in 1932 (p. 496) had no proof of this and declared that if S2 should turn out to contain S3 he would fall back on S1, which he knew to be weaker than S2. Although the normal system T was in existence from Feys 1937, it did not become well known until Sobociński 1953 proved it to be equivalent to the system M of von Wright 1951; and when Barcan 1946 appeared it was Lewis's S2 that she took as the basic modal system. The possible-worlds semantics for S2 was first presented in Kripke 1965. (Fuller details on all these matters may be found in Part III of Hughes and Cresswell 1968.) Despite being Lewis's fallback position, S1 does not have an intuitive semantics, and the present paper discusses the failure of several attempts to give it one. Thus the primary interest of S1 now is its link with the pioneers of modal logic, and it is in that spirit that I offer it in honor of perhaps the only surviving one of their number.

between worlds rather than between worlds and neighborhoods. Unfortunately, in one way or another, these semantics turn out to be defective, and in the second part of the paper, I shall discuss the 'impossibility' semantics presented in Girle 1975 and claimed to characterize S1 (or $S1^0$) in Schotch 1984 and Sylvan 1989, and the semantics involving a three-place relation presented on p. 151f of Zeman 1973.

I shall use the terminology and notation of IML except for the use of $\wedge$ for conjunction. (See especially the definition of S2-validity in IML p. 276.) An S1 model is a triple $\langle R, R^*, V\rangle$ in which R is a relation that is reflexive in its domain (i.e., it is what is called a *quasireflexive* relation in IML, p. 275.) Let $W = \mathfrak{F}R$ and let $N = \mathfrak{D}R$, so that R is reflexive in N. Let $O = W\text{-}N$. R^* is a 'neighborhood' relation, whose domain is included in W; that is, its first argument is a world and its second a set of worlds. For $w \in W$ and $A, B \subseteq W$, R^* satisfies the condition

(1) If wR^*A and wR^*B then $A \cup B \neq W$.

(We shall only be interested in wR^*A, wR^*B when $w \in O$, so (1) could be restricted to such w's without loss.)

V satisfies the usual requirements for atomic wff and truth functors. For L we have

(2) If $w \in N$, then $V(L\alpha,w) = 1$ if $V(\alpha,w') = 1$ for all w' such that wRw'. Otherwise $V(L\alpha,w) = 0$.

(3) If $w \notin N$, then $V(L\alpha,w) = 1$ if (a) $V(\alpha,w) = 1$ and (b) $wR^*|\alpha|$, where $|\alpha| = \{w \in W\colon V(\alpha,w) = 1\}$. Otherwise $V(L\alpha,w) = 0$.

This semantics is a mixture of the familiar relational semantics for ordinary modal logic, and the neighborhood semantics for more general intensional logics. It would certainly be possible to use a single neighborhood relation to cover both (2) and (3) but that would fail to capture the similarity between S1 and S2. S2 models are simply S1 models in which R^* is empty, and of course T models are S1 models in which $N = \mathfrak{D}R = W$. Note that R^* in this paper differs from the R' of Cresswell 1972 (p. 491). In that semantics $M\alpha$ was false at w when $wR'|\alpha|$. (For $S1^0$ models requirement (3a) is removed, as is the quasireflexiveness of R. Alternatively, (a) could always be omitted from (3) and, for S1, it be required that if wR^*A then $w \in A$.) A modal wff α is *S1-valid* iff $V(\alpha,w) = 1$ for every $w \in N$ in every S1-model $\langle R,R^*,V\rangle$

THEOREM 1. S1 is sound.

I shall use Lemmon's axiomatization of S1 as presented on p. 246f of IML and will refer to the rule of substitution of proved strict equivalents by 'Eq'. Lemmon's basis for S1 consists of the following axioms:

(4) every PC-tautology
(5) $Lp \supset p$
(6) $(L(p \supset q) \wedge L(q \supset r)) \supset L(p \supset r)$

together with their necessitated versions, and the rules of uniform substitution, Eq and material detachment. (For $S1^0$ omit (5).) First note that if $V(\alpha, w) = 1$ for every $w \in W$, then both α and $L\alpha$ are S1 valid. It is therefore sufficient to show that $V(\alpha,w) = 1$ for all $w \in W$, where α is one of (4)–(6). Axioms (4) and (5) are trivial. For (6) we consider two cases. If $w \in N$ and $V(L(p \supset r), w) = 0$ there is some $w' \in W$, such that wRw', $V(p, w') = 1$ and $V(r, w') = 0$. But if $V(L(p \supset q), w) = 1$, this means that $V(q, w') = 1$, and if $V(L(q \supset r), w) = 1$, this in turn means that $V(r, w') = 1$, contradicting its falsity there. So (6) is true at all normal worlds. Suppose $w \notin N$. Then if (6) fails at w we have $V(L(p \supset q), w) = 1$. So $wR^*|p \supset q|$. But $|p \supset q| \cup |q \supset r| = W$; for take $w^* \notin |p \supset q|$; i.e., $V(p \supset q, w^*) = 0$. Then $V(q, w^*) = 0$. But then $V(q \supset r, w^*) = 1$; i.e., $w^* \in |q \supset r|$. So not $wR^*|q \supset r|$ and so $V(L(q \supset r), w) = 0$, contradicting the assumed falsity of (6) at w. For Eq we note that if $(\alpha = \beta)$ is valid then $|\alpha| = |\beta|$ and so $|L\alpha| = |L\beta|$ and so $(L\alpha = L\beta)$ is also valid. Eq then follows by induction on the construction of wff. Material detachment and uniform substitution are clearly validity preserving. This proves Theorem 1.

We now move on to the completeness of S1. Using $\top$ as an abbreviation for $p \supset p$ we may first note that the following hold in S1:

LEMMA 2. If $\vdash L\alpha$ then

(i) $\vdash L\top \prec \alpha$
(ii) $\vdash L\top \prec L\alpha$

For (i) observe that $p \equiv (\top \supset p)$ is a PC-tautology, and so $\vdash p = (\top \supset p)$, and so $\vdash \alpha = (\top \supset \alpha)$. So, by Eq, $\vdash L\alpha = L(\top \supset \alpha)$, and so, since $\vdash L\alpha$, $\vdash \top \prec \alpha$. But $\vdash L\top \prec \top$, since $p \supset \top$ is a PC-tautology. So $\vdash L\top \prec \alpha$. For (ii) observe that $\vdash \top \prec \alpha$, and that $\alpha \supset \top$ is an instance of a PC tautology. So $\vdash \top = \alpha$ and so, by Eq, $\vdash L\top = L\alpha$. So $\vdash L\top \prec L\alpha$.

LEMMA 3. $\vdash \alpha \rightarrow \vdash L\top \prec \alpha$

Proof: If α is an instance of an axiom then $\vdash L\alpha$ and so, by Lemma 2(i) $\vdash L\top \prec \alpha$. And if α is the necessitation of an axiom then by Lemma 2(ii) $\vdash L\top \prec \alpha$. If α is obtained from β by Eq by the substitution of γ for δ, then $L\top \prec \alpha$ can be obtained from $L\top \prec \beta$ by the same substitution. If α is obtained from β and $\beta \supset \alpha$ then the S1 theorem

(7) $((p \prec q) \wedge (p \prec (q \supset r)) \supset (p \prec r)$

will enable $L\top \prec \alpha$ to be derived from $L\top \prec \beta$ and $L\top \prec (\beta \supset \alpha)$.

LEMMA 4. $\vdash \alpha, \vdash \sim L\top \prec \alpha \rightarrow \vdash L\alpha$

For Lemma 4, use the following instance of Axiom (6):

(8) $\vdash ((\sim q \prec p) \wedge (p \prec q)) \supset (\sim q \prec q)$

to get

(9) $\vdash ((p \prec q) \wedge (\sim p \prec q)) \supset Lq$

Then with $L\top/p$, and α/q Lemma 4 follows from Lemma 3.

The worlds of the canonical model for S1 are as follows:

(10) $w \in N$ iff w is a maximal S1-consistent set of wff.
(11) $w \in O$ iff w is a maximal PC-consistent set of wff such that
  (i) $\{\sim L\alpha: \alpha \notin w\}$ is S1-consistent
  (ii) $L\top \notin w$

(11i) ensures that $L\alpha \supset \alpha \in w$ for any wff α, since no set containing $\sim L(L\alpha \supset \alpha)$ is S1 consistent.

(12) wRw' iff $w \in N$ and $w' \in W$ and $L^-(w) \subseteq w'$

(Following Hughes and Cresswell 1984, p. 21, $L^-(w)$ is defined as $\{\alpha: L\alpha \in w\}$)

In what follows let $\|\alpha\| = \{w \in W: \alpha \in w\}$, while, given a particular model, $|\alpha|$ will, as before, refer to $\{w \in W: V(\alpha, w) = 1\}$.

(13) wR^*A iff $w \in W\text{-}N$ and there exists a wff α such that $A = \|\alpha\|$ and $L\alpha \in w$.

Finally, where p is any propositional variable:

(14) $V(p, w) = 1$ iff $p \in w$

LEMMA 5. R is quasireflexive with N its domain and W its field.

To prove lemma 5 we have to show two things, that R is reflexive in N, and that $W = \mathcal{F}R$. If $w \in N$ then, since $\vdash L\alpha \supset \alpha$ for every wff α, $L^-(w) \subseteq w$ and so wRw. So R is reflexive in N, and by (12) $\mathcal{D}R \subseteq N$. So, also, if $w \in N$ then $w \in \mathcal{F}R$. If $w \notin N$, then by (11i) $\{\sim L\alpha: \alpha \notin w\}$ is S1-consistent, and therefore there is some $w' \in N$ such that $\{\sim L\alpha: \alpha \notin w\} \subseteq w'$ so if $L\alpha \in w'$ then $\alpha \in w$, and thus $w'Rw$, and so $w \in \mathcal{F}R$. So $W \subseteq \mathcal{F}R$, and from (12) $\mathcal{F}R \subseteq W$.

THEOREM 6. $V(\alpha, w) = 1$ iff $\alpha \in w$

The proof is by induction on the construction of α with the only nontrivial case being the induction step for L. First suppose that $L\alpha \in w$. If $w \in N$ and

wRw' then $\alpha \in w'$, so $V(\alpha, w') = 1$ and so $V(L\alpha, w) = 1$. If $w \in O$ then by (13) $wR^*\|\alpha\|$, and by (11i) $\alpha \in w$. So, by the induction hypothesis, $wR^*|\alpha|$ and $V(\alpha, w) = 1$, and so $V(L\alpha, w) = 1$.

Suppose $L\alpha \notin w$ and $w \in N$. It will be sufficient to show that $L^{-}(w) \cup \{\sim\alpha\}$ has an extension in N or in O. If neither then

(i) For $\beta_1, \ldots, \beta_n$ with $L\beta_1, \ldots, L\beta_n \in w$ we have $\vdash (\beta_1 \wedge \ldots \wedge \beta_n) \supset \alpha$

(ii) For $\gamma_1, \ldots, \gamma_k$ with $L\gamma_i, \ldots, L\gamma_k \in w$ we have that $(\sim L\top \wedge \gamma_i \wedge \ldots \wedge \gamma_k) \supset \alpha$ is a PC-tautology

Note that, where w^* is any maximal PC-consistent extension of $L^{-}(w)$ then $\{\sim L\gamma: \gamma \notin w^*\} \subseteq w$, and so is S1-consistent, since $w \in N$. Thus any such w^* satisfies (11i). Let $\delta = (\beta_1 \wedge \ldots \wedge \beta_n \wedge \gamma_1 \wedge \ldots \wedge \gamma_k)$. Then, since $w \in N$, $L\delta \in w$. Further from (i) $\vdash \delta \supset \alpha$ and from (ii) $\sim L\top \supset (\delta \supset \alpha)$ is a PC-tautology, and so $\vdash \sim L\top \dashv (\delta \supset \alpha)$. So by Lemma 4, $\vdash L(\delta \supset \alpha)$ and so $\vdash L\delta \supset L\alpha$. So since $L\delta \in w$, $L\alpha \in w$ contrary to hypothesis.

Suppose $L\alpha \notin w \in O$. We need to show that there is no β such that $\|\alpha\| = \|\beta\|$ and $L\beta \in w$. If $\|\alpha\| = \|\beta\|$ then $\vdash \alpha = \beta$, for if not then for some $w^* \in N$, $\alpha = \beta \notin w^*$. So $(L(\alpha \supset \beta) \wedge L(\beta \supset \alpha)) \notin w^*$. But from the proof of the theorem for normal worlds this means there is some w' such that w^*Rw' and $\alpha \in w'$ and $\beta \notin w'$, or $\beta \in w'$ and $\alpha \notin w'$. In either case $\|\alpha\| \neq \|\beta\|$. Since

$$\vdash \alpha = \beta$$

then by Eq

$$\vdash L\alpha = L\beta$$

and so, by (12) and Lemma 5, for every $w' \in W$,

$$L\beta \supset L\alpha \in w'.$$

So, since $L\alpha \notin w$ then $L\beta \notin w$. This completes the proof of Theorem 6.

THEOREM 7. $\langle R, R^*, V\rangle$ is an S1-model.

Given Lemma 5 all that is necessary is to prove:

LEMMA 8. R^* satisfies (1).

To prove Lemma 8 suppose that wR^*A and wR^*B. Then $w \in W\text{-}N$ and there exist wff α and β such that $A = \|\alpha\|$, $B = \|\beta\|$, $L\alpha \in w$, $L\beta \in w$. It is sufficient to show that $\|\alpha\| \cup \|\beta\| \neq W$. Suppose, for reductio, that $\|\alpha\| \cup \|\beta\| = W$. Then $\|\alpha \supset \beta\| = \|\beta\|$, and so, as in the proof of Theorem 6,

(15) $\vdash (\alpha \supset \beta) = \beta$

and so, by Eq,

(16) $\vdash L(\alpha \supset \beta) = L\beta$

So, by (12) and Lemma 5, for all $w' \in W$

(17) $L\beta \supset L(\alpha \supset \beta) \in w'$

so since $L\beta \in w$, then

(18) $L(\alpha \supset \beta) \in w$.

But note that from

(19) $\vdash (L({\sim}p \supset p) \wedge L(p \supset (p \supset q))) \prec L({\sim}p \supset (p \supset q))$

we have

(20) $\vdash (Lp \wedge L(p \supset q)) \prec L\top$

and so for all $w' \in W$

(21) $(L\alpha \wedge L(\alpha \supset \beta)) \supset L\top \in w'$

and so, since $L\alpha$ and $L(\alpha \supset \beta) \in w$

(22) $L\top \in w$,

contradicting (11ii).

The proof for $S1^0$ proceeds mutatis mutandis. (Kit Fine has an unpublished proof along similar lines.)

The neighborhood semantics for S1 is not particularly elegant, but in my view is the best possible, and confirms the opinion I expressed in note 3 on p. 199 of Cresswell 1967 that S1 is "a very silly system," and deserving of attention only on historical grounds. A recent revival of interest in S1 has been motivated by the idea that in nonnormal worlds L means 'it is impossible that'. This semantics was used in Girle 1975 to separate S1 and S2, which indeed it does. Girle did not explicitly claim nor purport to prove completeness for the semantics; but such claims are made in Schotch 1984, p. 89f (for $S1^0$) and in Sylvan 1989. Since these claims are intended to provide S1 and $S1^0$ with a plausible interpretation, there would seem to be some merit in showing that neither system is complete with respect to the impossibility semantics. I shall first establish that the following is not a theorem of S1

(23) $(Lp \wedge Lq) \prec L(p \wedge q)$

Let

$$W = \{w_0, w_1, w_2, w_3\}$$

$$R = \{\langle w_0, w_0\rangle, \langle w_0, w_1\rangle, \langle w_0, w_2\rangle, \langle w_0, w_3\rangle\} \quad (\text{so } N = \{w_0\})$$

$$R^* = \{\langle w_1, \{w_1, w_2\}\rangle, \langle w_1, \{w_1, w_3\}\rangle\}$$

R^* satisfies the restriction since $w_0 \notin \{w_1, w_2\} \cup \{w_1, w_3\}$.

Let $|p| = \{w_1, w_2\}$, $|q| = \{w_1, w_3\}$. Then by (3) $V(Lp, w_1) = 1$, $V(Lq, w_1) = 1$. But $|p \wedge q| = \{w_1\}$ and so not $w_1R^*|p \wedge q|$ and so $V(L(p \wedge q), w_1) = 0$. So $V((Lp \wedge Lq) \supset L(p \wedge q), w_1) = 0$. So $V((Lp \wedge Lq) \prec L(p \wedge q), w_0) = 0$.

Formula (23) turns out to be valid according to the impossibility semantics, showing that S1 is not complete for that semantics. A *Girle – Schotch – Sylvan (GSS) model* (for S1) is a quadruple $\langle W,N,R,V\rangle$ in which $\emptyset \neq N \subseteq W$, $R \subseteq W^2$ is reflexive in N, and V satisfies the following:

(24) For $w \in N$, $V(L\alpha, w) = 1$ if $V(\alpha, w') = 1$ for all w' such that wRw'. Otherwise $V(L\alpha, w) = 0$.
(25) If $w \notin N$, $V(L\alpha, w) = 1$ if (a) $V(\alpha, w) = 1$ and (b) $V(\alpha, w') = 0$ for every w' such that wRw'. Otherwise $V(L\alpha, w) = 0$.

For $S1^0$ R is not required to be reflexive, and clause (a) can be dropped from (25). Validity is defined as truth in every model for every $w \in N$. I shall show that (23) is valid in a GSS model. It is sufficient to show that

(26) $(Lp \wedge Lq) \supset L(p \wedge q)$

is true in all worlds. For $w \in N$ the result is standard, so consider $w \in W\text{-}N$. If (26) fails at w then $V(Lp, w) = 1$ and $V(Lq, w) = 1$. So first, by (25a) $V(p, w) = 1$ and $V(q, w) = 1$. So $V(p \wedge q, w) = 1$. And by (25b) $V(p, w') = 0$ for every w' such that wRw', and so $V(p \wedge q, w') = 0$ for every w' such that wRw', so $V(L(p \wedge q), w) = 1$ as required.

Because of the failure of (23) in S1 there will be a nonnormal world w in its canonical model that will contain $L\alpha$ and $L\beta$ but not $L(\alpha \wedge \beta)$. This will prevent there being a world w' containing $\alpha \wedge \beta$ with the property that $\gamma \notin w'$ for every $L\gamma \in w$, as required in Sylvan's completeness proof on p. 31. Schotch's completeness proof proceeds by giving a GSS model equivalent to a neighborhood model, involving the method of Segerberg 1971, p. 20f. However, Segerberg's proof relies on having (26) in every world, and the absence of (23) from S1 prevents this from being guaranteed. While there can be no proof that *no* relational semantics will work – because it is impossible to set limits on the kind of nonstandard evaluation rule that might be used – it is difficult to see how the extremely weak constraint imposed by the neighborhood models could be expressed in a relational structure.

Given that S1 is *not* complete for the impossibility semantics, the question arises of what system *is* characterized by it. In fact it is S1 +

(27) $(Lp \wedge Lq) \prec L(p \vee q)$

and

(28) $({\sim}L\top \wedge Lp \wedge q) \prec L(p \wedge q)$

Call this system S1^{+}. Note that Lemma 4 still holds in S1^{+}. From (27) and (28) we may prove the following

LEMMA 9. $\vdash \alpha \prec (\beta_1 \vee \ldots \vee \beta_n \vee \gamma_1 \vee \ldots \vee \gamma_k)$
$\rightarrow \vdash ({\sim}L\top \wedge L\beta_1 \wedge \ldots \wedge L\beta_n \wedge L\gamma_1 \wedge \ldots \wedge L\gamma_k \wedge \alpha) \prec L\alpha$

In this lemma I shall abbreviate $(\beta_1 \vee \ldots \vee \beta_n \vee \gamma_1 \vee \ldots \vee \gamma_k)$ to η. First note that (27) generalizes to

(29) $(Lp_1 \wedge \ldots \wedge Lp_{n+k}) \prec L(p_1 \vee \ldots \vee p_{n+k})$

From this we have

(30) $(L\beta_1 \wedge \ldots \wedge L\beta_n \wedge L\gamma_1 \wedge \ldots \wedge L\gamma_k) \prec L\eta$

Now $(\alpha \supset \eta) \equiv (\alpha \supset (\alpha \wedge \eta))$ is a PC-tautology and so

(31) $\vdash (\alpha \supset \eta) = (\alpha \supset (\alpha \wedge \eta))$

So, by Eq

$$\vdash (\alpha \prec \eta) = (\alpha \prec (\alpha \wedge \eta))$$

So, from the hypothesis of the lemma,

(32) $\vdash \alpha \prec (\alpha \wedge \eta)$

Further, $(\alpha \wedge \eta) \supset \alpha$ is a PC-tautology and so

(33) $\vdash (\alpha \wedge \eta) \prec \alpha$

And so, from (32) and (33),

(34) $\vdash \alpha = (\alpha \wedge \eta)$

and so, by Eq,

(35) $\vdash L\alpha = L(\alpha \wedge \eta)$

Now $\vdash (p \prec q) \supset ((r \wedge p \wedge s) \prec (r \wedge q \wedge s))$. So from (30) by material detachment,

$$({\sim}L\top \wedge L\beta_1 \wedge \ldots \wedge L\beta_n \wedge L\gamma_1 \wedge \ldots \wedge L\gamma_k \wedge \alpha) \prec ({\sim}L\top \wedge L\eta \wedge \alpha)$$

but by (28)

$$\vdash (\sim L\top \wedge L\eta \wedge \alpha) \prec L(\alpha \wedge \eta)$$

so by (35) and Eq

$$\vdash (\sim L\top \wedge L\eta \wedge \alpha) \prec L\alpha$$

So, by (6),

$$\vdash (\sim L\top \wedge L\beta_1 \wedge \ldots \wedge L\beta_n \wedge L\gamma_1 \wedge \ldots \wedge L\gamma_k \wedge \alpha) \prec L\alpha.$$

This concludes the proof of Lemma 9.

The completeness proof now proceeds as for S1 except that the members of N are maximal consistent in $S1^+$ and R is extended so that for $w \in O$, wRw' iff for every $L\alpha \in w$, $\alpha \notin w'$. R^* of course plays no role.

THEOREM 10. $V(\alpha,w) = 1$ iff $\alpha \in w$.

The proof proceeds exactly as in Theorem 6 except for the case where $L\alpha \notin w \in O$. If $\alpha \notin w$ then we are done; so suppose $\alpha \in w$. It is sufficient to show that

(i) $\{\alpha\} \cup \{\sim\beta\colon L\beta \in w\}$

is $S1^+$-consistent or

(ii) $\{\sim L\top, \alpha\} \cup \{\sim\beta\colon L\beta \in w\} \cup \{\gamma\colon \vdash L\gamma\}$

is PC consistent. (i) ensures that α will be true in a normal world accessible from w, and (ii) ensures that α will be true in a nonnormal world accessible from w. The sufficiency of (ii) is perhaps less obvious, since we need to know that any maximal PC-consistent extension w^* of (ii) has the property that $\{\sim L\delta\colon \delta \notin w^*\}$ is $S1^+$ consistent. Suppose it were not. Then, for some $\delta_1, \ldots, \delta_h \notin w^*$ we would have

$$\vdash L\delta_1 \vee \ldots \vee L\delta_h$$

Let $\delta = (\delta_1 \vee \ldots \vee \delta_n)$. Then $\delta \notin w^*$. But by

$$\vdash (Lp \vee Lq) \supset L(p \vee q)$$

we have

$$\vdash L\delta$$

contradicting the fact that $\{\gamma\colon \vdash L\gamma\} \subseteq w^*$. So suppose that neither (i) nor (ii) holds. Then for some $\beta_1, \ldots, \beta_n$, with $L\beta_1, \ldots, L\beta_n \in w$,

$$\vdash \alpha \supset (\beta_1 \vee \ldots \vee \beta_n)$$

and for some $\gamma_1, \ldots, \gamma_k$ with $L\gamma_1, \ldots, L\gamma_k \in w$ and some $\delta_1, \ldots, \delta_h$ with $\vdash L\delta_s, \ldots, \vdash L\delta_w$

$$\delta_1 \supset (\ldots (\delta_h \supset ({\sim}L\top \supset (\alpha \supset (\gamma_1 \vee \ldots \vee \gamma_k)))))$$

is a PC-tautology. So

$$\vdash \delta_1 ⥽ (\ldots (\delta_h \supset ({\sim}L\top \supset (\alpha \supset (\gamma_1 \vee \ldots \vee \gamma_k)))))$$

So, by repeated application of

$$\vdash L(p \supset q) \supset (Lp \supset Lq)$$

$$\vdash L\delta_1 \supset (\ldots (L\delta_h \supset ({\sim}L\top ⥽ (\alpha \supset (\gamma_1 \vee \ldots \vee \gamma_k)))))$$

And so by successive applications of material detachment

$$\vdash {\sim}L\top ⥽ (\alpha \supset (\gamma_1 \vee \ldots \vee \gamma_k)).$$

So by Lemma 4

$$\vdash \alpha ⥽ (\beta_1 \vee \ldots \vee \beta_n \vee \gamma_1 \vee \ldots \vee \gamma_k)$$

So, by Lemma 9,

$$\vdash ({\sim}L\top \wedge L\beta_1 \wedge \ldots \wedge L\beta_n \wedge L\gamma_1 \wedge \ldots \wedge L\gamma_k \wedge \alpha) ⥽ L\alpha.$$

But ${\sim}L\top, L\beta_1, \ldots, L\beta_n, L\gamma_1, \ldots, L\gamma_k, \alpha \in w$, and so $L\alpha \in w$, contradicting the assumed consistency of w. This proves Theorem 10.

The model used to falsify (23) will also falsify (27) and (28) so neither is a theorem of S1. If we add $\langle w_1, \{w_1\}\rangle$ to R^* we will validate (28) but not (27), and if we add $\langle w_1, \{w_1, w_2, w_3\}\rangle$ we will validate (27) but not (28), showing that these are independent. From (28) in S1 we may derive (23).

The other relational semantics that has been developed for S1 is found on p. 151f of Zeman 1973. Zeman proposes a three-place relation whose first argument is filled by what might, for convenience, be called a time. Where U is such a relation, the rule for L in a nonnormal world w is that $L\alpha$ is true in w iff

(36) there is a time t such that α is true in every w' such that $Utww'$, provided that α is false in at least one world.

Zeman's semantics does not validate (23) and should be compared with the semantics used in Schotch and Jennings 1981 and elsewhere to study nonaggregative modal logics, in which even (26) fails. However, this semantics does validate

(37) $L(p \wedge q) ⥽ (Lp \vee Lq)$

and would validate the proper S2 axiom

(38) $L(p \wedge q) \dashv Lp$

were it not for the proviso in (36) that α be false somewhere.

(39) $L(p \wedge q) \supset (Lp \vee Lq)$

holds in every nonnormal world according to Zeman's semantics but not according to the S1 semantics presented earlier. The latter is established simply by changing the S1 model used to falsify (23) so that

$$R^* = \{\langle w_1, \{w_1\}\rangle\}$$

For the former note first that if $p \wedge q$ is false somewhere then at least one of p and q must be false somewhere. And if, for some t, $p \wedge q$ is true at every world accessible at t from w then so are both p and q. So either Lp or Lq will be true at w. (This is so even without the condition that Zeman imposes on U to obtain the effect of (1).) Zeman on pp. 155–60 provides a completeness proof for S1 (and $S1^0$) using semantic tableaux. As far as I can tell this proof is correct, but the semantics used is in fact the version of the neighborhood semantics found in Cresswell 1972. On p. 152, Zeman shows that each of his own models induces an equivalent neighborhood model, but the direction he needs is that every neighborhood model can be induced by one of his own models. That he does not prove, and the failure of (37) in S1 shows that it cannot be done.

The invalidity of (37) has other consequences. For (39) will be true in any world in which L is interpreted as 'it is possible that', and also in any world in which L is interpreted as 'it is possibly not that'. This means that in any semantics whose nonnormal worlds give L either the impossibility interpretation, or the possibility interpretation, or the possibly-not interpretation, the following wff will be valid

(40) $L\ (((Lp \wedge Lq) \supset L\ (p \wedge q)) \vee (L\ (r \wedge s) \supset (Lr \vee Ls)))$

That (40) is not a theorem of S1 may be seen by extending the model used to falsify (23) by letting $|r| = \{w_0, w_1, w_2\}$ and $|s| = \{w_1, w_2, w_3\}$. So no subsystem of S1 can be characterized by any class of models each of whose nonnormal worlds receives one of the interpretations just mentioned. For this reason none of the semantics offered in Sylvan 1989 for $S0.7^0$–S0.9 is correct, and his completeness proofs fail for the same reasons as his S1 proof. Since Sylvan's paper opens with such marvelous rhetoric it is perhaps a little sad that his semantics, which he calls "markedly superior" (to neighborhood semantics) should lack the only virtue that the latter possess, of being correct.

REFERENCES

Barcan, R. C. 1946. A Functional Calculus of First Order Based on Strict Implication. *Journal of Symbolic Logic* 11: 1–16.

Cresswell, M. J. 1967. The Interpretation of Some Lewis Systems of Modal Logic. *Australasian Journal of Philosophy* 45: 198–206.

1972. The Completeness of S1 and Some Related Systems. *Notre Dame Journal of Formal Logic* 13: 483–95.

Feys, R. 1937. Les logiques nouvelles des modalités. *Revue Néoscholastique de Philosophie* 40: 217–52.

Girle, R. A. 1975. S1 ≠ S0.9 *Notre Dame Journal of Formal Logic* 16: 339–44.

Hughes, G. E., and M. J. Cresswell. 1968. *An Introduction to Modal Logic* (IML). London: Methuen.

1984. *A Companion to Modal Logic*. London: Methuen.

Kripke, S. A. 1965. Semantical Analysis of Modal Logic II, Non-normal Modal Propositional Calculi. In J. W. Addison et al. (eds.), *The Theory of Models*. Amsterdam: North Holland, pp. 206–20.

Lewis, C. I. 1918. *A Survey of Symbolic Logic*. Berkeley: University of California Press (N.B.: The chapter on strict implication is not included in the 1961 Dover reprint.)

Lewis, C. I., and C. H. Langford. 1932. *Symbolic Logic*. New York: Dover Publications 1932, 2nd ed. 1961.

Schotch, P. K. 1984. Remarks on the Semantics of Non-normal Modal Logics. *Topoi* 3: 85–90.

Schotch, P. K., and R. E. Jennings. 1981. Non-Kripkean deontic logic. In R. Hilpinen (ed.), *New Studies in Deontic Logic*. Dordrecht: Reidel, pp. 149–62.

Segerberg, K. 1971. *An Essay in Classical Modal Logic*. Uppsala: Filosofiska Studier.

Sobociński, B. 1953. Note on a System of Feys–von Wright. *Journal of Computing Systems* 1:171–8.

Sylvan, R. 1989. Relational Semantics for All Lewis, Lemmon and Feys' Modal Logics, Most Notably for Systems between $S0.3^0$ and S1. *Journal of Non-classical Logic* 6: 19–40.

Wright, G. H. von. 1951. *An Essay in Modal Logic*. Amsterdam: North Holland.

Zeman, J. J. 1973. *Modal Logic: The Lewis Modal Systems*. Oxford: Clarendon Press.

4

A Problem in Possible-World Semantics

DAVID KAPLAN

I. Introduction

I believe that there is a problem in the conceptual/mathematical foundation of *possible-world semantics* (PWS) which threatens its use as a correct basis for doing the model theory of intensional languages, not of all intensional languages, but of some.[1] This is *not* the notorious problem that logically equivalent expressions have the same intension in PWS. It is a problem that arises even if we limit our intensional languages to notions for which substitution under logical equivalence is entirely correct.[2] I can't say that I fully understand the problem, though I have thought about it off and on for about fifteen years.

I call the problem conceptual/mathematical. Mathematically, it is comparable to, indeed it uses an argument based on the same mathematical foundation as, Russell's paradox, which displayed a contradiction in naive set theory. I cannot give my problem quite so complete a mathematical formulation, but I do think that it shows a serious difficulty in the naive foundations of PWS.

We may approach the problem by means of an analogy. Suppose that one proposed to conceive of the semantics of first-order quantification theory in terms of the method of quantifier elimination and truth-functional evaluation in successively larger finite domains, the logical truths being thought of as those sentences that transform into tautologies in every case.

In an age of Frege, Ruth Barcan Marcus's Russellian instincts, first expressed in the path-breaking "A Functional Calculus of First Order Based on Strict Implication," and reiterated in an influential symposium with Quine and Kripke, can be seen as opening the door to the flourishing international trade in quantified modalities. Though I was only a late convert to her Russellianism, I was early captivated by her remarkable intellect, scholarly range, and philosophical insight, always accompanied by her sprightly, ever good-humored charm. Ours has been a sustaining friendship for well over two decades. She continues to hold my admiration and great personal affection. ["A Functional Calculus of First Order Based on Strict Implication" is in the *Journal of Symbolic Logic* 11 (1946): 1–16. The symposium includes Marcus's "Modalities and Intensional Languages," *Synthese* 13 (1961), and "Discussion," *Synthese* 14 (1962), both reprinted in Ruth Marcus, *Modalities: Philosophical Essays* (Oxford: Oxford University Press, 1993).]

© Copyright by David Kaplan.

There is nothing incoherent or contradictory about such a method. It is just that because it excludes the infinite models, it does not capture the proper intuitive idea of *all* possibilities, which is the idea we aimed to explicate with our notion of logical truth. The problem with this method does not show up in monadic quantification theory. But it constrains the interpretation of non-logical *dyadic* predicates by making it impossible for there to be a relation satisfying the intuitively consistent sentence:

$$\forall x \exists y Rxy \wedge \forall xyz(Rxy \wedge Ryz \rightarrow Rxz) \wedge \forall x \sim Rxx$$

So here, in essence, is the problem in PWS. The possible-worlds methodology imposes constraints on the properties expressed by intensional constants, constraints that make it impossible for there to be properties satisfying certain seemingly benign conditions. It appears, therefore, that PWS does not encompass all the intuitive possibilities. The fact that not all possibilities are included seems to be an unintended consequence of the fundamental ideas we associate with PWS.

II. The Problem

There are difficult and profound philosophical issues regarding intensional notions. For example, some may subscribe to a metaphysics according to which all of the intensional supervenes on the extensional, that what is said, known, believed,[3] required, forbidden, desirable, necessary, or contingent is, in this metaphysical sense, supervenient on the distribution of earth, air, fire, and water. According to this view, once we have represented all the possible distributions of the elements, we have represented all possibilities.[4] Thus, each complete distribution of the elements either would necessitate it being desirable that all should be water or would necessitate it not being desirable that all should be water.

But if PWS is to serve for intensional *logic*, we should not build such metaphysical prejudices into it. We logicians strive to *serve* ideologies not to constrain them.[5] Thus, insofar as possible, our intensional logic should be neutral with respect to such issues. Concretely, this means that our lexicon should permit, in every grammatical category, nonsupervenient, nonlogical constants expressing arbitrary properties. For example, we would ordinarily expect our modal logic to permit the existence of a property such that for any set of individuals it is possible that the property apply to exactly those individuals.[6] As logicians, we would not *require* that there be such a property, for to do so would be to refute Hyperdeterminism – the view that *whatever is, is necessary* – according to which a property *could* not apply to what it *does* not apply to. To require this would be to meddle in metaphysics. But I can think of no plausible reason why logic itself should refute the existence of such a property.[7]

A somewhat weaker property, here expressed by the monadic predicate U, is one such that for any individual, x, it is possible that the property applies to x and only to x.

$$\forall x \Diamond \forall y (Uy \leftrightarrow y = x)$$

It takes ingenuity to think of an example of such a predicate[8] (more ingenuity when the domain of individuals is infinite), but again it seems plain that an adequate modal logic should not rule it out on *logical* grounds.

If we now consider propositions and their properties (as usually expressed by intensional sentential operators), the analog to the monadic predicate U would be a sentential operator, Q, such that for any proposition, p, it is possible that the property expressed by Q holds of p and only of p.

(A) $$\forall p \Diamond \forall q\, (Qp \leftrightarrow p = q)$$ [9]

Natural examples of operators, Q, that satisfy (A) are difficult to think of. Perhaps, for every proposition, it is possible that it and only it is *Queried.*[10] Perhaps not.[11] It shouldn't really matter. There may be no operator expressible in English which satisfies (A). Still, *logic* shouldn't rule it out.

In the usual possible worlds methodology,[12] a model for a sentence such as (A) has the following features.

(1) We are given a set W (of "possible worlds").
(2) If $w \in W$, ϕ is a formula, and f an assignment of values to variables, then f satisfies $\Diamond\phi$ in w iff f satisfies ϕ in some $w' \in W$

If the language is extended in the natural way to include propositional variables and nonlogical sentential operators, we add

(3) The propositional variables range over all subsets of W (the 'propositions').
(4) We are given an assignment, I, of intensions[13] to nonlogical constants. For the nonlogical sentential operator Q, $I(Q)$ assigns to each $w \in W$, a set of propositions (i.e., a family of subsets of W).
(5) If $w \in W$, ϕ is a formula, and f an assignment of values to variables, then f satisfies $Q\phi$ in w iff $\{w' \in W: f \text{ satisfies } \phi \text{ in } w'\} \in I(Q)\,(w)$.[14]

(A) has no models of this kind.[15]

III. Propositions

In PWS, wherein propositions just are (or are represented by) sets of possible worlds, there are plainly more propositions than possible worlds. But even on other natural conceptions of the relation between propositions and possible worlds, it seems odd for there to be a set of possible worlds not characterized by any proposition.

I take it that a (the) fundamental idea of PWS is to take possible worlds seriously, not necessarily by being a 'realist' about possible worlds, but by using possible worlds as a fundamental conceptual building block in the analysis of certain forms of speech.[16] There is an analogous and intimately related theory (or program), which we might call *possible-world metaphysics* (PWM), that attempts to use the same conceptual apparatus to analyze directly certain metaphysical notions (forms of speech aside). In both PWS and PWM it is traditional to introduce propositions as ancillary entities, definable in terms of possible worlds. There are, of course, other ways to proceed. One could, consistent with the possible-worlds apparatus, take the notion of proposition as a separate, primitive idea, and develop a theory about the relation between propositions and possible worlds, including the relation between a proposition and the set of possible worlds at which it holds.[17]

However, so long as the notion of a possible world remains fundamental, and the notion of a proposition is not regarded as utterly language dependent,[18] it is hard to conceive why there should be sets of possible worlds not characterizable by any proposition. Would these be sets of worlds so utterly disparate that they can only cohere by enumeration? So disparate that there is no sentence of any possible language of any possible life-form that characterizes them? So disparate that there is no property, no matter how elaborate or arcane, unique to just those worlds? And even then, why *not* characterize sets of worlds by 'enumeration'? If to each world there corresponds a proposition characterizing exactly it,[19] and if arbitrary (including infinite) disjunctions of propositions themselves form propositions (as they do in PWS), then all sets of possible worlds are characterized by propositions.

IV. Doing the Numbers

If propositions are entities, then our modal logic, in its premetaphysical purity, should accommodate a nonlogical sentential operator satisfying (*A*).[20] (*A*) tells us that there are at least as many 'possibilities' as there are propositions. If each 'possibility' corresponds to at least one possible world, then:

(1) There are at least as many possible worlds as propositions.

If possible worlds are entities, then for a variety of reasons:

(2) There are at least as many propositions as sets of possible worlds.

This cannot be.

V. Two Approaches

Without claiming complete understanding, let me try to explain one aspect of the underlying problem as I see it. We may distinguish those approaches to the model theory of PWS wherein the entities representing the possible worlds are thought of as constructed, for example as models for a given lexicon of nonlogical constants, from those approaches in which they are taken as *given* (as points if you will) and the model theory is done on them 'externally'.

Let us consider the first approach. Assume a simple lexicon of nonlogical, extensional constants and the one nonlogical intensional sentential operator *Q*. Each possible world (i.e., on this approach, each model) must assign an extension to each extensional constant, and a family of propositions to the operator. Hence, according to PWS, the model must assign to the operator a family of sets of models. But suppose that the model is itself a member of one of those sets.[21] How can a model assign a family of sets to a constant, if the model itself is a member of one of those sets? Unfoundedness yawns. On normal set-theoretic assumptions this is impossible.

So suppose that instead we assign to the operator a family of sets of models *of level 0* (i.e., models for just the extensional part of the language). This will avoid the unfoundedness difficulty. However, the operator will never hold of a sentence which contains it. No iteration. The operator will only be defined, so to speak, for sentences in the extensional part of the language. And the models containing this assignment will be models *of level 1*.

Models of level 1 seem to give rise to new possibilities beyond those of level 0. At level 0, there were only the possibilities of, say, different distributions of earth, air, fire, and water. Now there are different distributions of earth, air, fire, and water with it being queried whether all is earth, and there are the same distributions with it not being queried whether all is earth.[22] These new possibilities provide new propositions to query – for example, whether it is queried whether all is earth. The question whether this new proposition is queried is not already answered by the initial assignment of a family of sets of models of level 0 to the operator. It didn't arise at that time. The new propositions are represented by sets of models of level 1. Thus the assignment to the operator required to make it true that one of the new propositions is queried must take place in models of level 2. What we see here is the classification of both models and propositions into orders.

One benefit of the classification of propositions into orders is that it tames an intensional version of the Liar Paradox. Consider the sentence

S $$\forall p\,(Qp \rightarrow \sim p).$$

If the proposition expressed by S is the only proposition of which *Q* holds, then it is true iff it is false. Hence, given (A), a contradiction ensues.

Do not blame (A)! It is no more pernicious than the empirical assumption that the sentence engraved on Epimenides' tomb is 'The sentence engraved on Epimenides' tomb is false.' The problem lies not in the fact; it lies in our semantics.[23]

On the present approach, the argument fails because it ignores the hierarchical classification of propositions into orders. The argument requires the propositional variable in S to range over the very proposition expressed by S itself, which is not in accordance with the construction as (partially) described.

This process will never give us the big dividend of possible-world semantics: All iterations of all operators being settled at once. It is the commitment to a single fixed class of propositions, closed under application of intensional operators, that gives PWS its characteristic elegance in the treatment of iteration. It would not be PWS without it. Our hierarchical process cannot achieve that result.[24] The historically minded will remember that it was exactly the iteration principles that set Carnap to work trying to create new semantical methods, which ultimately gave us PWS. I note, as an aside, that Carnap's early work did not suffer from the present difficulty because he did not attempt to take account of nonlogical intensional constants.[25]

According to the second approach, a possible world is thought of not as a model, explicitly carrying limited information for a previously fixed lexicon, but as a point ('Let *W* be an arbitrary nonempty set') which implicitly carries information for all nonlogical constants. (Or perhaps as a Mundus Rasus on which information for arbitrary nonlogical constants can be writ.)

This approach either solves, or helps to conceal, the difficulty that appears so visible in the hierarchical results of the first approach.[26] This external method – advocated nowadays by almost everyone – is just to *assume* that all the possibilities are already out there, each represented by one of the points we call 'possible worlds'. It is this approach that is presupposed by Kuhn when he writes,

> But the point of possible worlds semantics is to interpret a new lexicon of intensional operators entirely in terms of the possible worlds that could have been stipulated before the new lexicon was added.[27]

That is, we have 'external' functions that tell us what the domain of individuals is for the already 'stipulated' possible world represented by each such 'point', what the assignment of values to each of the nonlogical constants is for each such 'point', etc. Kripke models ϕ built on a structure $\langle G, K, R \rangle$ are naturally taken to be of such a kind.[28]

I believe that the same difficulty, which shows up in the first approach by requiring hierarchy, reappears in this approach through constraints like that which makes (*A*) contradictory. However, most systems of PWS successfully avoid the difficulty, just as Carnap originally did, by avoiding arbitrary nonlogical constants and variables of intensional types.

VI. Closing

Using a theory known to have problems is problematic, but can be stimulating and useful. Naive set theory is like that. Although one doesn't see how to resolve all of the problems, there are areas in which one feels secure. Even on insecure terrain, we can often make good use of a theory. It may be all we have.

I don't feel secure with any nonlogical – perhaps I should say nonsupervening – intensional notions when they are interpreted by PWS. I especially worry about using it for propositional attitudes (even their E-versions), which are classic cases of nonlogical, arguably nonsupervenient, intensional operators. But I won't give up using PWS because I don't have a good alternative and because the problem is still somewhat mysterious.

I have tried to show that naive possible-world semantics leads to a kind of paradox just as naive set theory does, and by means of a similar argument. I also suspect that the ultimate lesson is somewhat the same, namely that the fundamental entities must be arranged in a never completed hierarchy and cannot be taken to be given all at once.[29] The foregoing considerations seem to suggest that an intensional logic that is suited to serve as a general foundation for the introduction of arbitrary nonlogical constants should ramify the intensional types and, in particular, the type of propositions. This was Russell's view. Few espouse it today.

Appendix: Logically Possible versus Metaphysically Possible Worlds

There is a second problem in PWS, as sometimes practiced, to which I wish briefly to call attention. It concerns the attempt to use a single set of intensional notions to simultaneously represent metaphysical and semantical ideas.

Consider, for example, the question of how many distinct, true propositions there are. Church takes up this question and resolves that *qua logician* he should not stipulate to there being more than one true proposition because, I believe, he thought that to do so would be to (negatively) prejudge metaphysical views such as that which I have called Hyperdeterminism.[30] But he also wishes to use his intensional logic to interpret the semantical theory of sense and denotation. Now if the propositions represent (potential) meanings, then Hyperdeterministic metaphysics should have no effect whatsoever on the *number* of propositions. Hyperdeterminism is a *metaphysical* thesis, not a thesis about meaning.[31] The correct Hyperdeterministic conclusion is that all of the (many) true propositions are necessary.

I do not deny that the metaphysical and the logical are interwoven. The surprising result[32] that the logically indeterminate sentence 'Hesperus is Phosphorus' expresses either a necessary or impossible proposition, violated the metaphysical innocence of logicians. But innocence is a relative thing. The metaphysical should not be confused with the logical. If Hyperdeterminism were to imply that there is only one true proposition, then not only would whatever is true be necessary (as would be expected), but *all* propositional operators would become truth functional. Even if this were the only metaphysically possible world, should 'it is desirable that' and 'it is undesirable that' become truth functional?[33] I think not!

A proper PWS framework for a language containing both *possibility* and *desirability* operators should, I believe, allow the logical to dominate the metaphysical. Not the reverse, as Church would have had it. This means that Hyperdeterminism or not, we must retain all the points (representing so-called possible worlds) needed to distinguish the propositions expressed by inequivalent sentences. However, the Hyperdeterminist must regard all but one of these points as representing what is not really possible, in a word, as representing unreal possible worlds. What then does the Hyperdeterminist who is also a possible-worlds Realist say about such points? At this late date, he is surely too sophisticated to fall into the trap leading to the conclusion that desirability becomes truth functional. And he has too much integrity to insist that all logically possible worlds are real even if they aren't really possible. It would be reasonable to take the view that real possible worlds correspond to some of the points, namely, those that are metaphysically possible. But this is not a Realist interpretation of PWS, it is only a Realist interpretation of *real* possibility.[34]

NOTES

1. An abstract of the argument appeared in the *Abstracts of Sections 5 and 12: 7th International Congress of Logic Methodology and Philosophy of Science.* The talk took place during the Congress in Salzburg, Austria, July 11–16, 1983. A more detailed version, on which the present paper is based, was delivered at the Nobel Symposium in Stockholm, Sweden, during the summer of 1986. The first airing of the argument was during the middle 70s. I also presented it as part of my John Locke lectures in Oxford during spring of 1980. The present version has benefited from the quick-response team of Kit Fine and Terence Parsons, who suggested many valuable improvements. The argument has been reported, and sometimes discussed, by M. J. Cresswell [''Quotational Theories of Propositional Attitudes,'' *Journal of Philosophical Logic* (1980); reprinted with updated references in M. J. Cresswell, *Semantical Essays: Possible Worlds and Their Rivals* (Dordrecht, Boston, and London: Kluwer Academic Publishers, 1988): 100–1], Martin Davies [*Meaning, Quantification, Necessity: Themes in Philosophical Logic* (London: Routledge & Kegan Paul, 1981): appendix 9], David Lewis [*On the Plurality of Worlds* (London and New York: Basil Blackwell, 1986): sec. 2.3

"More Worlds Than There Are?" pp. 104–8], and Michael Jubien ["Problems with Possible Worlds" in David F. Austin (ed.), *Philosophical Analysis* (Norwell, Mass.: Kluwer, 1988): 299–322].

2. To notions for which substitution under logical equivalence is not correct, such as 'she whispered that', we may associate an E-analog, 'she E-whispered that', obtained by closing the notion in question under logical equivalence. E-whispering is a rather exotic notion, not often encountered in practical life. Theoretically, however, it is undefective.
3. More property, E-said, E-known, or E-believed.
4. Vague as this is, it already raises delicate questions about whether it is the distribution of *particular* quantities of the elements (such as the very air that, in the actual world, currently fills my lungs) or simply a question of what amounts of what elements are where.
5. Except to valid argument, of course.
6. For example, *being headed for Heaven,* under the old free-will scenario.
7. Hyperdeterminism is my favorite among totally implausible metaphysical views. Ours is the best of all possible worlds because it is the *only* possible world. It is said that it was the rigors of Hyperdeterministic Puritanism from which the original settlers of Southern California were fleeing.
8. Perhaps any one of us could have been the sole winner of the superjackpot.
9. We are unused to seeing propositional expressions flanking '=', although we certainly understand the meaning when the propositional expressions are variables (assuming we understand the domain over which propositional variables range, as PWS assures us we do). In a standard PWS model for an S5 modal logic, in which accessibility is the universal relation among possible worlds, '$q = p$' can be expressed by '$\Box (q \leftrightarrow p)$'.
10. That is, it is asked whether it is the case that *p*. (More properly, it is E-asked.) Note that on this reading, *Q* is neither closed under consequence nor closed under implication. '*Q* (You saw my proof before you published yours)' implies neither '*Q* (You published your proof)' nor '*Q* (I showed you my proof before you published yours)'. If one insisted that $\Box (Q\phi \leftrightarrow Q \sim \phi)$, which I do not believe, it would not affect the substance of what follows, though (A) would take a slightly different form.
11. Here is an argument against (*A*) when *Q* is so interpreted. For a proposition to be queried, it must be expressed. So if not all propositions can be expressed, then not all can be queried. But it is natural to think that there are more propositions than can be expressed in any one language. For one reason, because it is natural to think that there are only a denumerable number of expressions in any one language, and it is reasonable to think that there are more propositions than that. If we were to accept infinite disjunctions of propositions as forming new propositions (as I am inclined to do, and as PWS does), then for any collection of natural numbers, there would be a proposition saying that my lucky number is among them. This already gives us more than a denumerable number of propositions.

    In defense of (*A*): For it to be *possible* that a given proposition is queried, there need not be some one, previously fixed, language in which it and all other propositions are expressible. On this score, for (*A*) to hold, it need only be the case that for each proposition, it be possible that it is expressed in *some* language, perhaps a different possible language for each proposition. Are there propositions that are not expressible in *any* possible language used by *any* possible beings?

This would be an interesting result. Could it be a result of *logic*?

It should also be recalled that the presence of indexicals in a language can, on some theories, generate a multitude of propositions from just a few words.

12. For a simple S5 modality.
13. In Carnap's sense.
14. A more familiar form for clauses (4) and (5) is to replace $\boldsymbol{I}(Q)$ in clause (4) with an 'accessibility' relation, $\boldsymbol{R}$, between possible worlds, and to replace the right-hand side of clause (5) with a definition in terms of $\boldsymbol{R}$. Clause (5) might then take the form, f satisfies $\boldsymbol{Q}\phi$ in w iff f satisfies ϕ in some [all] w' such that $w\boldsymbol{R}w'$. It is apparent that such clauses can be replaced by clauses of the form of (4) and (5). For example, for the preceding case[s] $\boldsymbol{I}(Q)$ would be defined as follows: If $\mathbf{P} \subseteq W$ (i.e., $\mathbf{P}$ is a proposition), then for all possible worlds w, $\mathbf{P} \in \boldsymbol{I}(Q)(w)$ iff for some w' such that $w\boldsymbol{R}w'$, $w' \in \mathbf{P}$ [for all w' such that $w\boldsymbol{R}w'$, $w' \in \mathbf{P}$].

    Note however that not every intensional operator that can be represented in the form of clauses (4) and (5) can be represented in the more familiar form of a definition in term of an accessibility relation. If there are n worlds, then there are $2^{(n^2)}$ accessibility relations. There are only a denumerable number of definitions, so there are $\omega \times 2^{(n^2)}$ assignments that can be represented in the traditional way. But there are $n \times 2^{(2^n)}$ assignments that can be represented in the form of clauses (4) and (5). The latter is larger than the former when n is infinite.
15. The proof, which I will not spell out here, having already overdosed on cardinality arguments, is the obvious application of Cantor's Theorem.
16. An alternative, not implausible, way to regard the formal model theory associated with PWS is as a "mere" mechanism for articulating and verifying logical relations among sentences of the object language (and perhaps the meaningfulness of certain syntactic forms, such as quantification in), without any commitment to deeper conceptual or metaphysical foundations. On this alternative, the proof of PWS is in the accuracy and the didactic assistance it provides in predicting logical relations among (and perhaps meaningfulness of) object-language sentences containing modal or other intensional expressions. It is these object-language logical relations (and meaningfulness claims) that are subject to independent verification through intuition, philosophical argument, or whatever other means seem appropriate. Such intuitions would concern the notions of *possibility* and *necessity*, not the notion of a *possible world*. This is what I call "*not* taking possible worlds seriously."
17. I rather favor a version of this approach: taking propositions, properties, etc., to be fundamental 'intensional' (in a vague sense) entities, whose exact structure and individuation we do not yet fully understand, and regarding the Carnapian intensions of PWM (in the technical sense of functions from possible worlds to truth values, sets of individuals, and the like) as their 'extensions'. I know this is a confusing use of 'extension'.
18. Those cautious, antiseptic conceptions of proposition that are closely tied to expressibility in some fixed language seem to me to be in odd alliance with the robust, "Get metaphysical!" theology of so many (not all) PWS partisans. Neither the language dependent nor the metaphysical conceptions trouble me on their own, just the odd alliance.
19. Could there be two worlds so utterly alike that they cannot be distinguished except 'numerically'?
20. Suppose K were a collection of possible worlds including all possible distributions of Earth, Air, Fire, and Water. (*A*) requires that there be possible worlds

beyond K. Any such world would differ from all members of K with respect to the interpretation of Q, but would duplicate a member of K with respect to the distribution of Earth, Air, Fire, and Water. But this is exactly what Supervenience denies. So Supervenience looks skeptically at (A), and ultimately judges it false for *metaphysical reasons*. Fair enough. I have no problem accommodating a metaphysical viewpoint according to which (A) is false. I believe that modal logic should accommodate even Hyperdeterminism. But PWS judges (A) false for *pre*-metaphysical reasons. Unfair!

21. As it would be if any true proposition had the property expressed by the operator. For example, if the operator were $\Diamond$.
22. The acceptance of both possibilities amounts to a denial of supervenience. As already stressed, our logic should not force the acceptance of metaphysical theses.
23. The foregoing is an alternative argument that PWS, as traditionally practiced, cannot accommodate (A). Fine and Parsons both urged presentation of this Cantor-free version of the argument. This version has the additional interest of demonstrating that an initially plausible strategy proposed by David Lewis – "to deny that just any set of worlds gives the content of some possible thought [Query]" (On the Plurality of Worlds, p. 106) – does not, in fact, succeed in rendering (A) "unexceptionable" (p. 106) within the PWS framework.
24. There is a natural way in which to construct models at level ω provided that the hierarchy is constructed in such a way as not to redefine Q for propositions for which it has already been defined. There are some technical subtleties in this since the universal proposition, for example, is represented by a different set at level 0 than it is at level 1, but it can be done by requiring that the new assignments *fit together* with earlier assignments in a certain definable way.

    At level ω we will get all iterations without propositional variables settled. But new propositions will still arise. Thus in the case in which propositional variables are involved, as in (A), the difficulty remains.
25. I see the origins of PWS in Carnap's semantical explorations of modal logic in "Modalities and Quantification," *Journal of Symbolic Logic* 11 (1946): 33–64, and *Meaning and Necessity* (Chicago: University of Chicago Press, 1947). In Carnap, the possible worlds are represented by State Descriptions, a variant of our first approach. I believe that the difficulty about iteration presently under discussion does infect Carnap's more general analysis in his "Replies and Systematic Expositions" [in P. Schilpp (ed.), *The Philosophy of Rudolf Carnap* (LaSalle, Ill.: Open Court 1963), pp. 889–905.
26. For those who have the faith (that all the possibilities are really out there, all at once), it solves it.
27. Thomas Kuhn, "Possible Worlds in the History of Science," Preprint (1986), p. 23.
28. Saul Kripke, "Semantical Considerations on Modal Logic," *Acta Philosophica Fennica: Modal and Many Valued Logics* (1963) 83–94.
29. At the turn of the Century, Russell ended *The Principles of Mathematics* (London: Allen & Unwin, 1903) with these words:

    > The totality of all logical objects, or of all propositions [or possible worlds?], involves, it would seem, a fundamental logical difficulty. What the complete solution of the difficulty may be, I have not succeeded in discovering; but as it affects the very foundations of reasoning, I earnestly commend the study of it to all students of logic.
30. Alonzo Church, "A Formalization of the Logic of Sense and Denotation," in P. Henle et al. (eds.), *Structure, Method, and Meaning: Essays in Honor of Henry M. Sheffer* (New York: Liberal Arts Press, 1951), p. 22.

31. Perhaps this is more apparent if we recall that Church chose, as a partial principle of individuation for propositions what he called ''Alternative Two'': Logically, equivalent sentences express the same proposition. Is it not clear that Hyperdeterminism should have no effect on whether two sentences are logically equivalent?
32. Due to Saul Kripke.
33. If you doubt that these notions are closed under logical equivalence, use E-desirable and E-undesirable.
34. Readers may think that I am speaking covertly of my respected friend David Lewis, but I am not.

5

Senses of Essence

KIT FINE

One may distinguish between the essential and accidental properties of an object. A property of an object is essential if it must have the property to be what it is; otherwise the property is accidental.

But what exactly is meant by this account? It has been common to give a further explanation in modal terms. A property is taken to be essential when it is necessary that the object have the property or, alternatively, when it is necessary that it have the property if it exist.

For reasons that I have already given in my paper "Essence and Modality," I doubt whether this or any other modal explanation of the notion can succeed. Indeed, I doubt whether there exists any explanation of the notion in fundamentally different terms. But this is not to deny the possibility of further clarification; and it is the aim of the present paper to provide it.

What I shall do is to distinguish some of the closely related ways in which the notion may be understood. This will be important for getting clearer both on which claims can be made with its help and on which concepts can be defined with its help. In particular, we shall see that several different senses of ontological dependence correspond to the different senses of essence. The task is also important for the purpose of developing a logic of essentialist reasoning; for most of the different senses of essence that we distinguish will make a difference to the resulting logic. My main concern in this paper has been with making the distinctions, and not with drawing out their implications; but I hope it is clear from the examples what some of these implications are.

1. Predicational and Sentential Forms

It is helpful to distinguish two different grammatical forms that essentialist claims can assume.

I should like to thank the many people who in numerous discussions have helped me. Ruth Chang read through the whole paper and made valuable comments. My debt to Ruth Barcan Marcus is evident; for she has perhaps done more than anyone, in her technical and philosophical work, to help make the notion of essence respectable. It is therefore with gratitude as well as affection that I dedicate this paper to her.

The first takes the concept of essence to be expressed by means of a predicate modifier. Thus given a predicate P (say 'thinks'), we may regard the expression L of essence as a device for forming the corresponding essentialist predicate $[LP]$ ('essentially thinks'). So to say that Socrates essentially thinks, we would apply this complex essentialist predicate to a term for Socrates.[1]

The second form embodies the concept of essence by means of a sentential, rather than a predicate, modifier. Thus to express the claim that Socrates essentially thinks, we would first form the sentence 'Socrates thinks' (to avoid issues of scope, let us here suppose that the name 'Socrates' refers directly to Socrates). We would then prefix the opeator 'It is true in virtue of the identity of Socrates that' to obtain the sentence 'It is true in virtue of the identity of Socrates that Socrates thinks'.[2] The symbol L for essence, on this account, has something like the syntactic status of 'believes'; it applies to a term t and a sentence ϕ and results in a sentence, which we may write in the form $L_t\phi$.

The preceeding examples involve only a single subject. But similar considerations apply to the cases in which there are several subjects. If we want to say that Socrates and Plato are essentially distinct, for example, we may do this either by attaching the predicate 'is essentially distinct' to the two subjects 'Socrates' and 'Plato' or by attaching the operator 'it is true in virtue of the identities of Socrates and Plato that' to the sentence 'Socrates is distinct from Plato'.[3]

The two forms of expression are not mere notational variants. For the predicational mode enables us to capture certain subtle distinctions that are lost under the sentential mode. Consider, for example, the following four claims:

(1) Socrates is essentially self-identical.
(2) Socrates is essentially identical to Socrates.
(3) Socrates is essentially such as to have Socrates identical to him.
(4) Socrates is essentially such as to have Socrates identical to Socrates.[4]

Under the predicational approach, we can distinguish between them. For the first ascribes essential self-identity ($[L\lambda x(x = x)]$) to Socrates; the second and third ascribe two forms of essential identity to Socrates ($[L\lambda x(x = \text{Socrates}]$) and $[L\lambda x(\text{Socrates} = x)]$) to Socrates; and the last ascribes the degenerate property of Socrates' self-identity ($[L\lambda x(\text{Socrates} = \text{Socrates})]$) to Socrates. However, under the sentential approach, these four claims collapse into the single claim:

(5) it is true in virtue of the identity of Socrates that Socrates is identical to Socrates (or, more formally, L_{Socrates} (Socrates = Socrates)).[5]

There is therefore no single way of translating statements of the sentential mode into those of the predicational mode. If one insists on a translation, then probably the most natural method is to let all of the occurrences of the given objects be relevant to the truth of the given proposition. Thus (5) would be taken to be equivalent to the "pure" form (1).[6]

But despite the greater expressive subtlety of the predicational form of expression, there are reasons for adopting the sentential form. Nothing is thereby lost in the essentialist information that can be conveyed. For, as we shall later see, the truth values of the impure claims can always be determined on the basis of the truth values of the pure claims and hence of the sentential mode formulations.

But something is also gained in the ease with which essentialist claims can be made and manipulated. For under the predicational approach, we must constantly adjust the content of those claims to the order and multiplicity of the arguments. Suppose, for example, that we wish to add Plato as an irrelevant argument to the claim that Socrates is essentially a man. Then we will have to go from Socrates essentially having the property of being a man to Socrates and Plato essentially bearing a certain relation, namely the one that holds between two objects when the first is a man. But under the sentential approach, we can simply pass from its being true in virtue of the identity of Socrates that Socrates is a man to this same thing being true in virtue of the identities of Socrates and Plato.[7]

Of course, this is not to say that either approach is more basic than the other. In fact, I am inclined to regard the predicational approach as fundamental. But in this, as in so many other cases, the most basic form of expression need not be the most convenient.

It is worth mentioning another, somewhat different, approach to the formulation of essentialist claims. It is an algebraic, rather than logical, in character. The essence of an object is treated as a single item; and the symbol *L* for essence is therefore taken to signify the function that takes each object into this item, its essence.

Of course, in the spirit of the predicational or sentential approaches, one might regard the essence of an object as the class of its essential properties or as the class of propositions that are true in virtue of what the object is; and the "parts" of the essence could then be taken to be subclasses of this class. But one is not required to think of the essence and its parts in such a way; and for those philosophers who think of essence as some part of "core," it might be better not to. Thus the algebraic approach provides a framework that is neutral between essence as core or content; and it should therefore be helpful both for understanding the conception in terms of core and for relating it to the conception in terms of content.[8]

2. Comparison with Analyticity

There is a systematic parallel between the notions of necessity and essence, on the one hand, and the notions of analyticity and meaning, on the other.

A necessary truth can be taken to be a proposition that is true in virtue of the identity of all objects; and, in the same way, an analytic truth can be taken to be a sentence that is true in virtue of the meaning of all terms (not that all objects or all items need be relevant). Essence, under the sentential construal, becomes a localized form of necessity; it is truth in virtue of the identity of some among the objects. In the same way, we may introduce a localized form of analyticity: truth in virtue of the meaning of some among the terms. Thus it is natural, under such a conception, to say that the sentence 'All bachelors are unmarried' is analytic in 'bachelor' (and perhaps the logical constants) but not in 'unmarried'.

Given that there is a predicational form of expression for essence, this leads us to look for its semantic counterpart. What corresponds to a property or a relation is a sentential scheme; and what corresponds to the essential possession of a relation is the analytic fulfillment of a scheme. So instead of saying that the sentence 'All bachelors are unmarried' is analytic in the term 'bachelor', we can say that the scheme 'All *X*'s are unmarried' is analytically fulfilled by that term.

We are thereby able to make subtle discriminations corresponding to those of the preceding section. Thus we can say that the scheme 'all *X*'s are *X*'s is analytically fulfilled by the term 'bachelor', that the scheme 'all *X*'s are *Y*'s' is analytically fulfilled by the terms 'bachelor' and 'bachelor'; and similarly for the other cases. We are also able to define meaning in strict analogy with the definition of essence. So just as the essence of an object can be identified with the class of its essential properties, the meaning of a term can be identified with the class of sentences that are analytic in the term, thereby providing a much more satisfactory account of meaning than the one in terms of synonymy classes.

The parallel between essence and meaning is further discussed in my paper "Essence and Modality" and will be further developed as we proceed.

3. Consequential/Constitutive Essence

Let us return to the topic of essence. Our concern now is not with how essentialist claims are to be expressed but with how, once made, they are to be construed.

Say that the property Q is a (logical) *consequence* of the properties P_1, P_2, . . . , or that they (logically) *imply* Q, if it is a logical truth, for any object, that it has the property Q whenever it has the properties P_1, P_2, ... Then the first issue is whether the essential properties of an object should be taken to be closed under logical consequence. Given that the object essentially has the properties P_1, P_2, . . . , and that these properties logically imply the property Q, should it be supposed that the object essentially has the property Q?[9]

We need here to distinguish between what one might call the *constitutive* and the *consequential* conceptions of essence. An essential property of an object is a constitutive part of the essence of that object if it is not had in virtue of being a consequence of some more basic essential properties of the object; and otherwise it is a consequential part of the essence. This distinction corresponds roughly to the traditional distinction between essence and propria.[10]

The constitutive essence is directly definitive of the object, but the consequential essence is only definitive through its connection with other properties. It is in principle possible that a logical consequence of a constitutive part of the essence of an object should itself be a constitutive part of the essence, but as a general rule this will not be the case. Consider Socrates, for example. His essence will, in part, be constituted by his being a man. But being a man or a mountain will merely be consequential upon, and not constitutive of, his essence.

A similar distinction can be drawn for the localized conception of analyticity. Thus being unmarried will be constitutive of the meaning of 'bachelor'; the sentence 'All bachelors are unmarried' will be constitutively analytic in 'bachelor' (and perhaps also in 'all' and 'are'). On the other hand, being unmarried or unhappy will be consequential on the meaning of 'bachelor'; the sentence 'All bachelors are unmarried or unhappy' will be only consequentially analytic in 'bachelor'.

The concept of constitutive essence, if it can be made out, is an important one. For it enables us to track the truth of an essentialist claim down to its source or, at least, to factor out the contribution made by logic. But it is also important for understanding the nature of the logical concepts and of logical truth. For one wants to say that it is true in virtue of the nature of identity and universality that every object is identical to itself; and one wants to define a logical truth as one that is true in virtue of the nature of the logical concepts. But such claims make no sense when logical closure is in force. For then every logical truth will be true in virtue of any objects or even in virtue of none.

It might be thought, however, that the whole idea of a logically inert conception of essence (or of analyticity) is incoherent. The difficulty arises from the essence of objects taken in combination. We want it to be true in virtue of the nature of Socrates and of disjunction that he is a man or a mountain. But how can this be true? Its truth must somehow derive from the respective natures of Socrates and of disjunction. But if these natures are propositional in form, then we will need to presuppose some notion of consequence in order to derive the nature of the combination from them. If, for example, the relevant nature of Socrates is given by the proposition that he is a man and that of disjunction by the proposition that Socrates is a man or a mountain if he is a man, then we will need to presuppose the rule of modus ponens in order to derive the desired conclusion that Socrates is a man or a mountain.

To overcome this difficulty, we should think of the nature of the logical concepts (or of the meaning of the logical terms) as being given, not by certain logical truths, but by certain logical inferences. Thus what properly belongs to the nature of disjunction is the inference from p to (p or q) rather than the fact that p implies (p or q). (Thus this is a case in which one might want to think of the nature of something as being nonpropositional in character). That Socrates is a man or a mountain will then follow from the nature of Socrates and of disjunction in the way in which a proposition is said to follow from certain propositions by means of certain rules. The concept of consequence is not presupposed but is already built into the rules.

Even though the foregoing objection fails, there is still considerable doubt as to how the concept of constitutive essence is to be understood. One problem is to know *where* to draw the line. It is essential to singleton Socrates that Socrates be its sole member. But is it essential to the set that Socrates be a member and that any two members be the same? Or is it essential to the set that Socrates be a member and that any two members be the same if Socrates is a member? Or has one or both of these formulations strayed too far from the "canonical" formulation to be part of the essence? A related problem is to know *how* to draw the line. Given the whole essence of an object, on what basis do we break it down into its component properties? And given one such component, on what basis do we choose between its logically equivalent forms?

Perhaps these questions can be answered. But until they are, it seems advisable to work as far as possible with the consequentialist notion.

4. Consequence

I want now to consider more carefully what might be meant by logical consequence.

One issue concerns the connection between essence and existence. Let us suppose that Socrates is essentially a man. Then it might be argued that existing is a logical consequence of being a man on the grounds that, for any object, it is a logical truth that it exists. So it follows by Closure that Socrates essentially exists, which is absurd.

The error of this argument lies in its assumption that it is a logical truth that Socrates exists. Perhaps there is a conception of logical truth according to which this is so. But if, as is most natural in the present context, we take a logical truth to be one that is true in virtue of the nature of the logical concepts, then surely it is not a logical truth that Socrates exists. For there is nothing in the nature of existence, or even of existence and Socrates taken together, which requires that Socrates should exist.

The details of such an approach to the logic of existence are already clear from the case of quantified modal logic and so need not be pursued here.

Another issue concerns the status of "extraneous objects." Consider the proposition that 2 = 2. It is a logical truth and hence a logical consequence of any proposition whatever. So under the consequentialist conception, it is true in virtue of the identity of any objects whatever that 2 = 2. But this may be considered unacceptable; for one may not want an object to figure in the essence of another unless it genuinely pertains to the nature of that other.

We may overcome this difficulty by restricting the objects that figure in the consequentialist essence of something to those that pertain to its nature. In the same way, one might, under the consequentialist conceptions of knowledge or of belief, restrict the objects that figure in the propositions known or believed by someone to those that appropriately lie within her ken. And one might even do something similar under the localized consequential conception of analyticity and not allow the sentence 'All triangles are triangles', for example, to be true in virtue of the meaning of 'bachelor'.

We seem to have an intuitive understanding of the relevant notion of pertinence in these cases. But when it comes to essence, the relevant notion of pertinence (or dependence) can be defined. For it is characteristic of the extraneous objects that they can be generalized away. Thus not only is it true in virtue of the identity of Socrates that 2 = 2 but also that, for any object x, $x = x$. The objects pertinent to a thing's essence can therefore be taken to be ones which cannot in this manner be generalized away.

Matters are perhaps otherwise with belief or knowledge or analyticity. For could not a person have merely logical beliefs or knowledge concerning an object within his ken and hence could not two people have the same consequentialist beliefs or knowledge and yet different objects within their ken? Or again, there would appear to be no reason why one should not define a term like 'bachelor', call it 'tachelor', as 'unmarried man if all triangles are triangles'. The same sentences would then be true in virtue of the meaning of 'bachelor' and 'tachelor' even though the pertinent terms, the ones by which they were defined, were different. In these other cases, therefore, the admission of the constrained conception may represent a real conceptual advance over the mere acceptance of the unconstrained conception.

By using the constrained consequentialist conceptions, it is possible to define the relevant notions of pertinence in a very simple way. In the case of essence, for example, we may say that y is pertinent to x, or that x depends upon y, if it is true in virtue of the identity of x that $y = y$. Using the other conceptions of essence, even the constitutive one, it is not possible to give such a simple definition; and, indeed, second-order methods must be employed. There is therefore a special advantage to be gained from using the constrained conception within a first-order account of essence, such as is to be found in my paper "Logic of Essence."

The final issue over consequence concerns Property Abstraction. According to this principle, there is, for any suitable condition, a property that is possessed by an object just in case it conforms to the condition. Now this principle is often taken to be a logical truth, though of second-order rather than first-order logic. But if it is, then the closure requirement leads to counterintuitive results.

For consider again the example of Socrates' being essentially a man. From this it follows by Property Abstraction and Closure that Socrates essentially has the property of being a man. But this is intuitively incorrect. For his nature no more requires his possessing this property, or even there being this property, than it requires his membership in some particular set.

To solve the difficulty, we must explain why Property Abstraction is not a logical truth; and we must also explain, regardless of the logical status of Abstraction, how we can resist the inference from Socrates being essentially a man to his essentially having the property of being a man. The second of these problems can be solved in terms of the familiar distinction between a property occurring in predicative or subject position. In the proposition that Socrates is a man, the property of being a man occurs in predicative position: It is predicated of Socrates. On the other hand, in the proposition that Socrates has the property of being a man, the property of being a man occurs in a subject (or argument) position: It is not predicated of Socrates; rather, the relation of "having" is predicated of Socrates and it. What makes it unacceptable to say that Socrates essentially has the property of being a man is the occurrence of the property as a subject rather than as a predicate. We are not merely appealing to the property in order to say how Socrates is; we are also explicitly saying how Socrates is related to the property.

In terms of this distinction we may also explain why Property Abstraction should not be regarded as a logical truth. For Abstraction, in the form that is bothersome to us, requires the occurrence of the property as a subject; we are saying of the property of being a man, for example, that it is possessed by an object just in case that object is a man. We should therefore only expect the principle to be true in virtue of the identity of the given property. Thus the fact that something has the property of being a man just in case it is a man will have its source in the nature of the property, and not simply in the nature of the logical concepts.

Our point of view on this matter makes it vital that, in the formulation of essentialist claims, we heed the distinction between predicate expressions, such as 'is a man', and property terms, such as 'being a man': The former introduce properties into predicative position; the latter introduce them into subject position. Real difficulties of formulation can arise when a property is introduced into subject position in order to say something about that same property in predicative position. Suppose, for example, that I say 'being a man is an essential property of Socrates'. This might be intended to convey that Socrates essentially has the property of being a man. But it might also be meant to *convey* that Socrates is essentially a man or, to be more exact, that the proposition that results from essentially predicating the property of being a man of Socrates is true. Thus even when there is reference to a property, the property may be "external," if I can put it that way, to the content of the underlying essentialist claim.[11]

5. Mediate/Immediate Essence

Let us grant that singleton Socrates essentially contains Socrates – that is, that it is true in virtue of the identity of the singleton that it contains Socrates. Let us also grant that Socrates is essentially a man – that is, that it is true in virtue of the identity of Socrates that he is a man. Does it follow that it is true in virtue of the identity of singleton Socrates that Socrates is a man? In other words, given that the set depends upon the man, is the essence of the latter included in that of the former?

This issue turns on the question of whether we take the essence of an object to be mediate or immediate. It is of the immediate nature or identity of singleton Socrates to contain Socrates and of the immediate nature of Socrates to be a man. But is is only of the mediate nature of singleton Socrates that it should contain something which is a man.

In general, the mediate nature of an object will incorporate the nature of all of the objects upon which it depends; the nature will in this sense be hereditary. But the immediate nature will only include that which has a direct bearing on the nature of the object and will exclude whatever belongs in virtue of the nature of the objects upon which the given object depends.

A similar distinction can be drawn within the semantic realm. Thus we may take the sentence 'All bachelors are men' to be immediately analytic in the term, 'bachelor' but take the sentence 'All bachelors are male' to be only mediately analytic in the term (granted that 'All men are male' is analytic in 'man').

The two notions of essence give rise to two notions of dependence. One object will *immediately depend* upon an another if it pertains to the immediate nature of the other, while one object will *mediately depend* upon another if it pertains to its mediate nature. Thus singleton singleton Socrates will immediately depend upon singleton Socrates but will only mediately depend upon Socrates. Similarly, the two notions of local analyticity give rise to two relations of semantic dependence, which will hold between two terms according as to whether the one can be either immediately or mediately defined by means of the other.

It is clear from the preceding characterization of the distinction how the mediate concept of essence (or of analyticity) is to be defined in terms of the immediate concept. For let $D(x)$ be the immediate definition or essence of x, for any object x. We may close the resulting system of definitions under the rule that $D(y)$ is included in $D(x)$ whenever y occurs in a property of $D(x)$ (and cannot be generalized away). The members of the resulting definitions will then be the mediately essential properties under the constitutive conception of essence; and their consequences will be the mediately essential properties under the consequential conception.

However, it does not appear to be possible to define immediate essence in terms of mediate essence. Certainly, the immediate relation of dependence cannot be defined in terms of the mediate relation. It will not do to take x to depend immediately upon y just in case x mediately depends upon y and there is no object z, distinct from x and y, such that x mediately depends upon z and z immediately depends upon y. For the set whose members are Socrates and singleton Socrates will immediately depend upon Socrates, even though it immediately depends upon singleton Socrates, which, in its turn, immediately depends upon Socrates. And nor does any other definition appear to be possible.[12]

A similar difficulty seems to stand in the way of defining immediate in terms of mediate essence. For surely we should allow that something might belong to both the immediate and mediate essence of an object (just as something might be both an immediate and a mediate dependent). But then given that it *could* be both, on what basis do we decide whether it *is* both? Suppose, for example, that it is of the essence of a particular chicken x to come from a particular chicken egg y and of the essence of y to be a chicken egg. Then presumably it is of the immediate essence of x to come from y. But is it of the immediate essence of x to come from a chicken egg?

Thus agreement on mediate essence does not secure agreement on immediate essence. But further disagreement on immediate essence tends, in practice, to be rare or inconsequential. Agreement on the essence of sets in terms of ancestral membership, for example, suggests a natural division of immediate essence in terms of membership. This point is significant for the substantive study of essence. For it means that we can, for the most part, confine our attention to what belongs to the mediate essence of things and let considerations of immediate essence ride on the result.

6. Reflexive Essence

Say that a property of an object is *self-related* if it is a property that involves the object itself. Thus being caused by God is a self-related property of God;

given a set S of antifoundational set theory whose sole member is itself, containing S as a member is a self-related property of S; and, for any object x, being identical to x is a self-related property of x.

Corresponding to any self-related property of an object is a reflexive property; this is a property that involves a reflexive as opposed to a direct relationship to the object in question. Thus corresponding to the property of being caused by God is the property of being self-caused; corresponding to the property of containing S is the property of containing itself; and corresponding to the property of being identical to x is the property of being self-identical. When a self-related property involves several occurrences of the given object, there will, of course, be different reflexive properties according as to which of the occurrences are made reflexive.

I want to consider two connected questions. First, does any object essentially have a self-related property? And second, given a corresponding pair of self-related and reflexive properties, what is the connection between the essential possession of the one and of the other?

There is a corresponding problem concerning the essential possession of a *relation*; and analogous questions arise within the semantic realm. For we may ask: Can a scheme involving a given term, say 'all X's are bachelors', be analytically fulfilled by that term?; and what is the connection between the fulfillment of such a scheme and the fulfillment of the corresponding reflexive scheme ('All X's are X' in the case at hand)?

It needs to be emphasized that these questions only arise under the predicational mode of expression, using either properties or schemes. For under the sentential approach, the distinction between the essential possession of a self-related property and of the corresponding reflexive property disappears, as does the analogous distinction between the analytic fulfillment of the self-related scheme and of the corresponding reflexive scheme.

It is plausible to maintain that when an object essentially has a given self-related property then it essentially has the corresponding reflexive property. Consider, for example, the case of the ill-founded set S; and let us grant that S essentially has the property of containing S. Then on what basis can one deny that S is essentially self-membered? Indeed, in merely affirming that the set S essentially contains S, one seems to be missing out on something that is indispensable to the essence of S, viz. that S is *self*-membered. After all, S is not a regular singleton. There is something in its nature that distinguishes it in kind from a regular singleton. And what could this be if not the property of being self-membered?

One would also like, if possible, to have a noncircular account of the essence of each object, one in which the identity of the object is not presupposed. If self-related properties are included in the essence to the exclusion of their reflexive counterparts, then it is not in general clear how a noncircular account is to be given. For the self-related properties would involve a circle and would lack any obvious basis in essential properties that do not. On the other hand, if the reflexive properties are included in the essence, circularity can be avoided, for the reflexive properties can be attributed to the essence on the basis of their reflexive counterparts.

The converse direction is not so clear. Given that an object essentially has a reflexive property, then does it essentially have the corresponding self-related property? Suppose first that essence is taken in the constitutive sense. Then it is plausible to suppose that the object will not have the corresponding self-related property. For how can a pair of self-related and reflexive properties both be constitutive of the essence of a given object? Surely the second is consequential upon the first and hence not properly constitutive of the essence.

Suppose now that essence is taken in the consequential sense. There would then appear to be only two sensible grounds for admitting a self-related property into the essence. On the one hand, the property might be admitted on the basis of the corresponding universal property. Thus we could say that Socrates is essentially a man or admiring of Socrates on the grounds that he is essentially a man or admiring of anyone.

On the other hand, the self-related property might be admitted on the basis of the corresponding reflexive property. The presence of a reflexive property in the essence would be taken to be a quite general ground for admitting all of the corresponding self-related properties.[13] Thus given that Socrates is essentially self-identical, it would follow that Socrates is essentially identical to Socrates or even essentially such that Socrates is identical to Socrates.

We might refuse to countenance either ground for admission. For under the assumption that an object cannot be self-dependent, the constrained consequentialist conception would exclude any self-related property from its essence. But even if we lift the constraints on consequence, we might still accept the first ground and not the second. An object could then only make a "trivial" appearance in its own essence.

Either of these two alternatives provides us with a perfectly coherent way of maintaining that an object x is essentially self-identical while denying that it is essentially identical to x [a possibility excluded by Wiggins (1976), p. 47, but endorsed by R. Barcan Marcus (1971)]. The object x would be essentially self-identical, since self-identity to x will not follow from the x-free properties in its constitutive essence.

Finally, one might accept both grounds for admission. We would then obtain a conception of essence that fails to make any of the aforementioned discriminations. It is the one that stands in the most straightforward connection to the sentential approach. For a proposition would be true in virtue of certain objects x, y, . . . just in case those objects essentially entered into the corresponding degenerate relation.[14]

Perhaps there are further grounds for discriminating among the different self-related properties. But it seems plausible that the essential attribution of such a property will never be autonomous; it will always be determined on the basis of the purely reflexive properties. Thus however the determination is made, no extra information about the essence will thereby be conveyed.

7. Reciprocal Essence

The issue discussed in the previous section concerned direct circularity. We wished to know whether the essence of an object can (ineliminably) involve that object itself. But there is also an issue of indirect circularity. We may wish to know whether there can be two objects each of whose essence involves the other: and similarly for longer cycles.

Consider the case of two interrelated fictional objects, for example. Could it be maintained that it was of the essence of Holmes to be admired by Watson (or, at least, to be admired by him in the Conan Doyle stories) and also of the essence of Watson to admire Holmes. Or again, given two points x and y in Euclidean space, could it be maintained that x is essentially distinct from y and y essentially distinct from x.[15]

It should be noted that the present issue, in contrast to the previous one, also rises under the sentential approach. For we may ask whether it can both be true in virtue of the nature of Holmes to be admired by Watson and be true in virtue of the nature of Watson to be admired by Holmes. An analogous question also arises in the semantic realm. For suppose I provide a "simultaneous" definition, let us say of 'term' and 'formula' for the predicate calculus with descriptions. There is then a sentence T defining 'term' in terms of 'formula' and a sentence F defining 'formula' in terms of 'term'. Do I want to say that T is analytic in 'formula' and that F is analytic in 'term'?

As with the issue over reflexivity, there is a way of avoiding the cycles if one wants. For instead of saying that it is essential to Holmes that he be admired by Watson and essential to Watson that he admire Holmes, we can say that it is essential to both Watson and Holmes, as a pair, that the one admire the other. And in the definitional case, instead of saying that F is analytic in 'term' and T is analytic in 'formula', we can say that each is analytic in the pair 'term' and 'formula'. Thus the fact that the items are related in the way they are may be seen to arise from their collective, rather than from their respective, nature or meaning. The apparently circular understanding of each, taken separately, is replaced by a simultaneous understanding of the two taken together.

It seems very reasonable that reciprocal essences should be forbidden under the constitutive conception; for how else could a noncircular account of the essence of things be given? But this leaves it open whether reciprocal essence should be allowed under a consequential conception. We may put the issue in terms of two ways of passing from the constitutive to the consequential essence. Suppose that it is part of the constitutive essence of x and y (in that order) that they stand in the relation R. Then do we regard that as a reason for supposing that x essentially stands in the relation R to y and y essentially stands in the relation R to x? And, in general, do we regard a relational essence as giving rise to the corresponding reciprocated essences?

We can go either way. On the one hand, we might insist that for an object essentially to have a certain property it must have that property solely in virtue of what it is, without regard for the identity of the associated objects. On the other hand, one might allow an object to essentially have a certain property when it has that property in virtue of the identity of that object in collaboration with the other objects.

Each of these conceptions of essence gives rise to its own distinctive notion of dependence. Under the first, each of two objects cannot both figure in the essence of the other; and hence dependence will be strict or one-way. Under the second, each of two objects will be capable of figuring in the essence of the other; and hence dependence may be reciprocal or two-way. Thus we may say that Watson and Holmes or that two points in Euclidean space are reciprocally dependent under the second conception, but not under the first.

Although the first of these conceptions is closer to the constitutive conception than the second, it is in some ways easier to work with the second. Its great advantage is that it allows us to express the essence of all things in monadic fashion as the essence of particular things. Thus instead of having to say that Watson and Holmes essentially enter into the relation of admiring, we can say that Watson essentially admires Holmes and that Holmes is essentially admired by Watson; the essential relationship "dissolves" into the component essential attributions.

8. The Essential Manifold

We come now to a notion of essence that involves a much more radical departure than the others from the notion with which we began. We have supposed that each object has a unique essence or definition, where this is something that may be identified either with the class of properties that it essentially has or with the class of propositions that are true in virtue of what it is. But we must now consider the possibility that the essence or definition might not be unique. Of course, such a possibility can arise in a relatively straightforward way. For the essences might be parts of a more comprehensive essence; and the differences between them might be ones of "formulation" rather than logical content. But what I have in mind are cases that do not arise in either of these ways: The essences are already comprehensive; and the differences are in logical content, not formulation.

How can this be? Consider an Aristotelian view of color. On this view, each color can be defined as the color of something having that color. Thus if the color is red, then red can be defined as the color of x, where x is any red thing. There are therefore different definitions of red, one for each red thing. These definitions have a different logical content, since the exemplars are different; and they are not plausibly regarded as parts of a more comprehensive definition.[16]

Indeed, even if it were legitimate to define red as the color of this and that, it seems wrong to view the definition of red as the color of this as somehow partial, merely part of a more comprehensive definition in terms of several things. For it is not as if the reference to additional exemplars further helped in fixing what red is.[17]

Another example, significant in the philosophy of language, concerns complex properties. It might be supposed that each complex property can be defined to be the result of abstracting on a corresponding proposition. Thus the complex property of not being wise can be defined to be the result of abstracting Socrates from the proposition that it is not the case that Socrates is wise. But any other person would do in place of Socrates here; and so we obtain several different definitions of the property, one for each possible subject. A further example concerns homogeneous (indefinitely divisible) stuff. Any quantity of such stuff can be defined to be the sum of component quantities. But there are many such definitions; and one stands out as being correct as opposed to the others.

Moreover, it is not really single definitions that will have alternatives but whole systems of definitions. Perhaps odd can be defined as not even; and even as not odd. But we do not, on pain of circularity, want to adopt both definitions.

Given the possibility of alternative definitions, we have to distinguish among three notions of essence where before we had only one. There is first the idea of essence as manifold. This is constituted by the range of alternative definitions. There are then the component essences corresponding to each of the definitions. And there is, finally, the common essence, which is constituted by what is common to all the component essences. Suppose that an object has the alternative definitions D_1, D_2, ... Then the component essences E_1, E_2, . . . of the object can be represented by the respective logical closures of D_1, D_2 . . . , the manifold essence by the class $\{E_1, E_2, \ldots\}$ of those components, and the common essence by the logical closure of the disjunction of E_1, E_2, ...[18]

The manifold essence of an object is unique but multifarious rather than propositional in character. The component essences are in general not unique but of a propositional character; and the common essence is both unique and propositional in character. The component essences may not be necessary to the thing; it is not necessary, for example, that red is the color of any particular red thing. But the common essence will be necessary to the thing; it will be necessary, for example, that red is the color of something. It is therefore the common essence that corresponds most closely to our previous undifferentiated notion of essence.

There are two ways of talking about essence, corresponding to the manifold and the common essence. Adopting the predicational form of expression, we may say that *x definitively* has a certain property if its having that property follows from some definition of the object, and we may say that it *essentially* has the property if its having that property follows from every possible definition of the object. Thus red will definitively be the color of any particular red thing, and it will essentially be the color of some red thing. A similar distinction can be drawn under the sentential form of expression, using '*x* can be so defined that' when the source of the truth lies in some definition and using '*x* must be so defined that' when the source lies in any definition.

The logic of the operator '*x* must be so defined that' is like that for $\Box$ (necessity). The logic of the operator '*x* can be so defined that' is neither like that for $\Box$ nor like that for $\Diamond$ (possibility). It is unlike $\Box$ in that it can correctly apply to two sentences without correctly applying to their conjunction; and it is unlike $\Diamond$ in that it can correctly apply to a disjunction without correctly applying to either disjunct. The closest modal counterpart to the operator is $\Diamond\Box$, with $\Diamond$ capturing the modal force of 'can' and $\Box$ that of 'defined'.

Similar distinctions can be made in the semantic realm. Thus the simple notion of being true in virtue of the meaning of a given term gives way to the notions of being true in virtue of some account of the meaning and of being true in virtue of any account. Similar distinctions may also have a place in moral philosophy. For under a view under which there can be alternative but equally legitimate systems of morality, we will want to distinguish between what is obligatory in the sense of being required by some moral system and what is obligatory in the sense of being required by any moral system.

The distinction, in the case of essence, is of great importance for the study of metaphysics. For an object's definitive properties need not be determined by its essential properties. It is conceivable, for example, that someone might think it is part of our understanding of a color that it have instances and yet not think that it must be understood by reference to any of those instances. Such a philosopher might then agree with our Aristotelian on the essential properties of red, but disagree on the definitive properties; for he would not allow that red was definitively the color of any red thing. Thus a full account of the nature of things must include their definitive, and not merely their essential, properties.

The distinction also corresponds to an important distinction between two forms of ontological dependence. For we may say that an object depends upon another either when it occurs (ineliminably) in the common essence of the other or when it occurs in a component essence of the other. In the one case, the given object *can* be defined in terms of another; and in the other case, it *must* be. Thus in the example of homogeneous stuff, we might plausibly maintain that a quantity of the stuff depends, in the first sense, on any of its component stuffs, but not in the second sense.

We may note, in conclusion, that hardly any of the preceding distinctions would make sense under the modal conception of essence. So even though it might initially appear to be a departure from rigor, it is only the liberation from the modal conception that has made the present clarification of the notion possible.

NOTES

1. There are many variants on this approach. A relatively minor one treats L as a copulative device. Thus given n terms $t_1, t_2, \ldots, t_n$ and an n-place predicate P, $Lt_1t_2 \ldots t_nP$ is a formula that, under the original notation, would have been written as $[LP]t_1t_2 \ldots t_n$.

   Under a more radical variant, the complex essentialist predicates are formed directly from an underlying formula. Thus given n variables $x_1, x_2, \ldots, x_n$ and a formula ϕ, $(Lx_1x_2 \ldots x_n)\phi$ is taken to be a complex essentialist predicate which, under the original notation, would have been written as $[L(\lambda x_1x_2 \ldots x_n)\phi]$, where $(\lambda x_1, x_2 \ldots x_n)\phi$ is the abstract for the relation of ϕ-ing.

   One can avoid the need for predicate modification or abstraction altogether by treating L as an operator that takes n singular terms $t_1, t_2 \ldots, t_n$, n variables x_1, $x_2 \ldots, x_n$ (which subsequently get bound), and a formula ϕ into a formula t_1t_2, $\ldots t_nLx_1\, x_2 \ldots x_n\phi$. The new formula expresses what would have been expressed under our previous notations by the application of the complex essentialist predicate $(Lx_1x_2 \ldots x_n)\phi$ or $[L(\lambda x_1x_2 \ldots x_n)\phi]$ to the terms $t_1, t_2 \ldots, t_n$. In a sense, this notation is the least committal of the three, since it does not require the postulation of any intermediate syntactic entity.

   Granted that there is a distinction between a term for a property and a predicate, still other variants can be obtained by using a property term in place of a predicate in the previous accounts.
2. I do not think of the subexpression 'the identity of Socrates' as a significant grammatical component of the operator 'it is true in virtue of the identity of Socrates that', just as I do not think of 'not' as a significant grammatical component of the operator 'it is not the case that.' In general, I shall use the terms 'essence' and 'identity' (and sometimes 'nature' as well) to convey the same underlying idea.
3. The general distinction between the two forms of expression is familiar from the literature on intensional logic. See especially Kaplan's "Opacity," Wiggins' "The De Re 'Must' . . . ," and Baldwin's "Wiggins and the De Re 'Must'."

   Just as with the predicational approach, there are many variants on the sentential approach. Two minor ones are (i) to treat L as a device that takes n terms t_1,

. . . , t_n into a one-place sentential operator $L_{t_1,\ldots,t_n}$, and (ii) to treat $L_{t_1,\ldots,t_n}$ as a predicate of propositions rather than as a sentential operator.

A more radical alternative, which greatly increases the expressive power of the language, is to index the operator L with a predicate expression F; $L_F\phi$ is then taken to mean that ϕ is true in virtue of the identity of the objects a_1, a_2, . . . that satisfy F. Thus what was formerly expressed as $L_{s,t}\phi$ would now be expressed by $L_{\lambda x(x=s \vee x=t)}\phi$. This is the approach adopted in my paper "The Logic of Essence."

4. Ruth Barcan Marcus distinguishes the modal analogues of the first three of the claims in her paper "Essentialism in Modal Logic."
5. Expressive parity can be restored if one gives the variables a literalist role (as described in Fine 1989) rather than a referential role. (1) to (4), for example, can be distinguished within the sentential mode as: $\exists x\ (x = s\ \&\ L_x\ (x = x))$; $\exists x\ (x = s\ \&\ L_x\ (x = s))$; $\exists x\ (x = s\ \&\ L_x\ (s = x))$; and $\exists x\ (x = s\ \&\ L_x\ (s = s))$. For the literalist character of the variable serves to link the subject of the essentialist attribution with its different positions in the embedded proposition. Expressive parity cannot properly be restored by complicating the embedded sentence. Thus $L_s\lambda x\ (x = x)s$ corresponds to $[L\lambda x\ (\lambda x\ (x = x)x)s$, and the like, and not to $[L\lambda x\ (x = x)]s$.
6. Certain technical difficulties arise in carrying out the proposed translation. Let us take $L_{s,t}\phi$ as a specimen statement to be translated. As a first step, note that (a) $L_{s,t}\phi$ is equivalent to a statement of the form $\exists x \exists y\ (x = s\ \&\ y = t\ \&\ L_{x,y}\phi)$, (b) descriptions can be eliminated from ϕ in the manner of Russell, and (c) when $\phi = \phi(u)$ contains a proper name u, $L_{x,y}\phi$ may be rewritten in the form $\exists z\ (z = u\ \&\ L_{x,y}\phi(z))$. Thus our problem reduces to that of translating $L_{x,y}\phi$, where ϕ is a formula whose only singular terms are variables.

   The obvious solution is to translate it as $((Lx,\ y)\phi)\ x,\ y$. But this will not do for three reasons: (i) x and y might be the same variable, in which case the notation $((Lx,\ y)\phi)$ will either be ill-formed or not have the intended sense; (ii) there may be other variables in ϕ that denote one of the objects denoted by x or y; (iii) the property expressed by the predicates in ϕ might involve the objects denoted by x and y.

   The first difficulty is readily overcome. For $L_{x,x}\phi$ is equivalent to $L_x\phi$; and hence, by conflating identical variables, we may suppose that the indexing variables are all distinct.

   The second difficulty may be overcome by being explicit about identities. Suppose that ϕ contains one free variable z other than x and y. Write ϕ in the form $\phi\ (x,\ y,\ z)$. Then $L_{x,y}\phi$ is equivalent to the disjunction of $x = y\ \&\ y = z\ \&\ L_x\phi\ (x,\ x,\ x)$, $x = y\ \&\ y \neq z\ \&\ L_x\ (x,\ x,\ z)$, $x = z\ \&\ x \neq y\ \&\ L_{x,y}\phi\ (x,\ y,\ x)$, $y = z\ \&\ x \neq y\ \&\ L_{x,y}\phi\ (x,\ y,\ y)$, and $x \neq y\ \&\ y \neq z\ \&\ x \neq z\ \&\ L_{x,y}\phi\ (x,\ y,\ z)$. In this way, we get a reduction to the case in which the free variables of $L_{x,y}\phi$ can be assumed to denote distinct objects.

   The difficulty posed by (iii) is not so straightforward. We could just stipulate that the primitive predicates of our language are to be 'pure' in the sense of expressing properties that do not involve objects. But if impure primitive predicates are allowed, then we must have some means of expressing a property as a function of the objects that it contains. We must be able to write P as $P_{x,y}$ with the dependence upon the possible constituents x and y made explicit.

   A related translation problem is discussed by Kaplan in Appendix B of his paper *Opacity*. He asks (p. 271):

   How do we understand

(22) **Bel** (y is taller than z)
when "y" and "z" have the same value? Do we understand it as:
(23) [(***Bel*** yz) y is taller than z] yz
or as:
(24) [(***Bel*** y) y is taller than y] y?

To which question, he replies:

Since we do not, in general, know when distinct variables have the same value, it seems that an understanding of (22) which would be uniform for all values of "y" and "z" should favor (23) over (24).

This leads him to "articulate" the free variables governed by **Bel** rather than to identify them, as in the proposal above. However, there is no reason why he should not use something like the solution to difficulty (ii) above to achieve a uniform translation that would render (22) equivalent to (24).

The logic governing the connection between the de re and de dicto forms of essence is somewhat different from that governing the connection between the corresponding *de re* and *de dicto* forms of belief; and this makes it all the more pressing to pursue a general line of investigation into the relationship between the two forms of expression.

7. Another example concerns the definition of absolute necessity, i.e., of truth in virtue of the identity of all objects. This is readily achieved under the sentential approach, at least if we use a predicate suffix; for $\Box\phi$ can be defined as $L_{\lambda y(y=y)}\phi$. But it is much more difficult to see how it is to be done under the predicational approach. For we must somehow convert the universe of objects into a sequence and then extend the predicational tie to arbitrary sequences. We might also note that the sentential approach enables one to fit the logic of essence into the orthodox framework of modal logic.
8. Another interesting application of the algebraic approach is to the question of understanding the informal account of what it is to be an essential property. One wants to say that the essential properties of an object are those that it must have if it is to be what it is. But what is meant by the phrase "it is to be what it is"? We might, using the algebraic approach, take it to express the propositional essence $L(x)$ of the object x – i.e., the maximal proposition true in virtue of what x is. Thus what is being said is that this proposition necessarily implies that the object x has the property in question.
9. Strictly speaking, it is better if the question is not posed in terms of properties, since this then raises the irrelevant issue of their existence and identity. Let us use the neutral form of notation $xLy\phi$ from note 2 in place of the committal form $(Ly)\phi x$. Then our question is whether $xLy\phi_1$, $xLy\phi_2$, ... imply $xLy\psi$ given that $(\phi_1 \,\&\, \phi_2 \,\&\, \ldots \rightarrow \psi)$ is a logical truth. However, for ease of expression, I continue to talk of properties in the text.
10. We shall later consider various ways in which the conception of consequence, and hence the corresponding conception of essence, can be modified; and there may be further ways still that are worthy of consideration.
11. There may be a special reason, on our approach, for *requiring* an external reference to properties in the formulation of essentialist claims. The question is whether in the basic notation xLP for such claims we should take P to express or to refer to a property. Suppose we wish to say that x essentially has some property. Then this is straightforward on the second option; we can say $(\exists P)$ $[xLP]$. But it gives rise to a possible problem on the first option. For suppose that we cannot quantify with respect to expressions in predicative position. Then we must say something like $(\exists P)$ $[xL\lambda x\ (x \text{ has } P)]$, where $\lambda x\ (x \text{ has } P)$ is the

complex predicate formed from the property term P. But this does not have the required meaning; for the essential property it attributes to x is that of having the property P.

12. Footnote 16 of my "Aristotle on Matter" constitutes a kind of proof of this fact.
13. No new essential properties not involving the object itself would thereby be introduced. To see this, first note that it suffices to add the property of being identical to the given object to the essence. For suppose that $\lambda x\phi\ (x)$ is the constitutive essence of x_0 and that ψ is a logical consequence of $\phi\ (x)$. Then ψ' will be a logical consequence of $(\phi\ (x)\ \&\ x = x_0)$, where ψ' is the result of replacing any occurrences of x in ψ with x_0.

    Now suppose that ψ is free of x_0 and a logical consequence of $(\phi(x)\ \&\ x = x_0)$. Then ψ is a logical consequence of $(\phi\ (x)\ \&\ x = x)$ and hence of $\phi\ (x)$.
14. Corresponding distinctions could also be made out for the localized conception of analyticity. Thus we might deny that the scheme 'All X's are bachelors' is analytically fulfilled by the term 'bachelor' on the grounds that the substitution of 'unmarried man' for X does not turn it into a logical truth. Or, on the contrary, we might take the scheme to be analytically fulfilled by the term on the grounds that the corresponding reflexive scheme 'All X's are X' is analytically fulfilled by the term.
15. Husserl was very much exercised by the possibility of reciprocal dependence; and his work (especially the third part of the *Logical Investigations*) provides other possible examples of the phenomenon. The case typified by Watson and Holmes is discussed at greater length in Fine 1982.
16. Rogers Albritton has pointed out to me that it might be though odd that the object x was not essentially red. Thus knowing the proposed definition of red and the nature of the object x would not enable one to ascertain the nature of the color. This problem, if it is one, might be overcome by requiring the object x to be a 'red'–something which by its nature is red just as a cube, one might suppose, is by its nature a cube.
17. I am assuming, of course, that the additional exemplars do not help in fixing the span of red. If multiple exemplars are allowed, then we cannot simply take a complete definition to be one whose content is maximal; for the definitions of red in terms of a single thing are complete and yet contained in the definitions in terms of several things. Another, perhaps more plausible, example of the same phenomenon is the definition of a point as the limit of an infinite sequence of regions; any such definition will be properly contained in, and will also properly contain, another such definition. In this case, unlike the color one, no natural class of minimal definitions can be singled out.
18. The disjunction is not quite right. Suppose that $x_1, x_1, \ldots$ are in fact all of the red things. Then it will be common to all of the definitions of red that it is the color of one of $x_1, x_2, \ldots$. But this makes the common essence dependent, in an unintended way, upon the way the world is. What we want is that the common essence should *necessarily* be common to the component essences. But in what sense of necessity? I suggest: in the sense yielded by the common essence of the object. There is the danger of circularity here; but it can be avoided. For it is reasonable to assume that the fragmentation of essence ceases at the second level. Thus even if red can have alternative definitions, there will not be alternative systems of such alternatives.

REFERENCES

Baldwin, J. 1979. Wiggins and the De Re Must, In *Papers on Language and Logic*. Keele: Keele University Library.

Barcan Marcus, R. 1967. Essentialism in Modal Logic. *Noûs* 1, no. 1: 91–6.

1971. Essential Attribution. *Journal of Philosophy* 67, no. 1.

Fine, K. 1982. The Problem of Non-existents I–Internalism. *Topoi* 1: 97–140.

1989. The Problem of De Re Modality. In J. Almog, J. Perry, and H. Wettstein (eds.), *Themes from Kaplan*. Oxford: Oxford University Press.

1992. Aristotle on Matter. *Mind* 101, no. 401, pp. 35–57.

1994. Essence and Modality. In J. Tomberlin (ed.), *Philosophical Perspectives*. Atascadero, Calif: Ridgeview, pp. 1–16.

Forthcoming. The Logic of Essence. *Journal of Philosophical Logic*.

Kaplan, D. 1986. Opacity. In E. Halnz and P. A. Schilpp (eds.), *The Philosophy of W. V. Quine*. La Salle, Ill.: Open Court.

Wiggins, D. 1976. The De Re 'Must': A Note on the Logical Form of Essentialist Claims. In G. Evans and J. McDowell. (eds.). Oxford: Clarendon Press.

6

Structuralism and the Concept of Set

CHARLES PARSONS

The term 'structuralism' with reference to mathematics has perhaps a basic meaning as referring to the idea that mathematics is particularly concerned with structures or structure. My concern is a more specific one, with what I call the structuralist view of mathematical *objects*. In its general lines, it is familiar: It holds that reference to mathematical objects is always in the context of some structure, and that the objects involved have no more to them than can be expressed in terms of the basic relations of the structure. The idea is well expressed by Michael Resnik:

> In mathematics, I claim, we do not have objects with an 'internal' composition arranged in structures, we have only structures. The objects of mathematics, that is, the entities which our mathematical constants and quantifiers denote, are structureless points or positions in structures. As positions in structures, they have no identity or features outside of a structure.[1]

Such a view has a long history, going back to the late nineteenth century. Through this history runs an important ambiguity, which implies a dilemma. Certain ways of stating structuralist views of particular kinds of mathematical objects rely on the concept of set or equivalent concepts, but that seems not to accord with the general intent that the structuralist view should be a general conception of mathematical objects, sets included. That one might rely on the concept of set is suggested by the notion of structure itself, which as a mathematical concept (for example in model theory) is usually understood in set-theoretic terms. Different structuralist views have either differed with respect to whether they were prepared to rely on the concept of set or been

This paper is not so closely connected with Ruth Barcan Marcus's work as one might ideally wish. But I should point out that modality plays a role here and has figured more prominently in other work of mine on the concept of set. And in connecting these two domains Ruth was a pioneer, in her paper "Classes, Collections, and Individuals," *American Philosophical Quarterly* 11 (1974): 227–32. That is certainly not the only area in which I have learned from her.

Versions of this paper have been presented to several audiences, most recently at a meeting in Budapest in May 1993 of the Académie Internationale de Philosophie des Sciences. I am grateful for the comments of hearers not all of whom I remember individually. Saunders Mac Lane, Rosemarie Rheinwald, and Wilfried Sieg deserve special mention.

unclear on the point.[2] Some have appealed to second-order logic, which puts the question back to the well-known one how much the concept of set needs to be relied on in understanding second-order logic. But now to rely on a set-theoretic conception of structure seems to make sets an exception to one's structuralism and to vitiate the claim of structuralism to be a *general* account of mathematical objects, or at least of pure mathematical objects such as elements of the various number systems and set-theoretic constructions from them, pure sets, and the like.

In a previous paper, I have discussed the problem of formulating a general structuralist view of mathematical objects.[3] But although I claim to have stated a viable version of the view, in its application to higher set theory my version does not avoid a problem that any version has to face. Attempts to use the structuralist idea to eliminate reference to mathematical objects of various kinds have a number of difficulties, but there is one in particular on which such a program founders when applied to set theory. Such attempts still require that one make it out that a structure of the kind required is in some way *possible.* In the previous paper I argue that in the present state of knowledge, this possibility can be made out for higher set theory only by domains of pure abstract objects, of which pure mathematical objects are a paradigmatic instance.[4] Does this imply that, to establish the possibility of the kind of structure described in axiomatic set theory, we need to appeal to a system consisting of *sets?* Such an admission would threaten a structuralist view of set theory with circularity and perhaps force us to embrace the horn of our dilemma that takes sets as an exception to a structuralist view of mathematical objects.[5] The purpose of the present paper is to explore this objection to a structuralist view of the objects of set theory and to argue that it is not in the end sustained.

I

In stating the objection, I said "threaten" and "perhaps" because what would defeat the structuralist view is not reliance on a system of sets per se, but such reliance where what we are appealing to is something more than a domain of 'objects' related in a relation called 'membership' satisfying conditions that can then be stated in the language of set theory. Where individuals (*Urelemente*) are involved, to be sure, the answer can be yes for a relatively trivial reason, that the individuals might be identifiable independently and not be preserved in an isomorphic structure. It seems rather clear to me, however, that that is just an element of generality in the notion of the 'universe of sets': It is not a unique structure because set theory does not determine what is an individual, how many there are, or what structure they have other than belonging to the sets they do. A structuralist view of set theory with individuals should treat the specific domain of individuals as belonging to the realization of the structure rather than as fixed.[6]

There is, however, a more serious reason arising from this consideration for thinking a structuralist view of set theory to be false. It comes from intuitions about sets of a more general ontological character. For example, there is the conception of a set as a totality 'constituted' by its elements, so that it stands in some kind of ontological dependence on its elements, but not vice versa. This would give to the membership relation some additional content, still very abstract but recognizably more than a pure structuralism would admit. The 'iterative conception of set' (a conception, not primarily of what a set is, but of what the 'universe of sets' is, and therefore of what sets there are) is usually explained with appeal to such general ontological conceptions, although also with appeal to ideas that cannot in the end be taken literally. One might put the objection in the following way: The 'objective reality' of the concept of set which on the structuralist view needs to be made out, cannot be made out except by describing structures consisting of *sets* (or some functional equivalent), where a set may still be ontologically impoverished compared to a concrete object but is still more than structurally determined.[7] We might call this view the 'ontological' conception of the objects of set theory, as opposed to the structuralist one.

It should be stressed at the outset that the objection to structuralism based on this ontological point rests on an epistemological one. The structuralist conception of set theory is rejected on the ground that without a richer conception of sets we cannot know that set theories describe coherent possibilities. Although it then becomes natural to interpret set theory by that richer conception, it does not follow that a structuralist interpretation of set theory is not possible at all and will not sometimes be found at work in mathematics, as it obviously will be in the study of models of set theory.

The claim made in the objection is that in order to convince ourselves of the possibility of rich enough structures to interpret set theory, we need to appeal to certain 'ontological' marks of the concept of set. That in fact such ontological conceptions are appealed to in justifying the axioms of set theory can readily be seen by examining well-known expositions on the subject. The question is then whether this appeal in fact has the implications the objection claims.

One consideration that might make us doubtful is the following: Differing characterizations of what a set is have been advanced in the history of the subject, and it appears that at different points in the development of the subject different characterizations are appealed to. One could distinguish between the idea of collection, the idea of plurality, and the idea of extension.[8] By a collection I mean an object that consists of its elements; it is typically thought of as in some way 'formed' from them; genetically, it no doubt derives from products of collecting. Because the formation of collections needs to be thought of in an abstract and general way for the notion even to begin to be adequate for set theory, it is not difficult to cast the notion into disrepute. But it has survived, perhaps because it motivates something central to the avoidance of paradox in axiomatic set theory, the priority of the elements of a set to the set: Thought of as a collection, a set is formed from objects that are already 'available' or 'given'. Pluralities I think of as given by plural constructions: the cats in my house, the prime numbers. One may doubt that there is a clear distinction between the idea of collection and that of plurality; one might think of them as rather two different ways of arriving at the same conception. Or one might resist the identification only by purism about the plural: It is already regimentation to think of the cats in my house as *a* plurality, but if one resists such regimentation, plural descriptions like the preceding simple examples needn't be taken to designate objects. The fact that this question arises in a way it doesn't for the idea of collection shows that there is an initial distinction. The last conception, that of extension, is possibly the clearest, because of its explanation by Frege: Roughly, extensions would be objects correlated with predicates, such that coextensive predicates have the same extension. It is an 'extensionalized' version of the very traditional idea of a property or attribute, which is so entrenched in our ordinary thought that even well-trained philosophers take for granted that predicates designate properties, in spite of the vulnerability to Russell's paradox of an unrestricted version of that view.[9]

The view I would defend of the role of these conceptions of what a set is is that they do play an indispensable role in the development of the concept but that none of them is adequate to set theory as it now stands (or even as it was constructed in Cantor's time). They certainly appear at important points in the history of set theory, especially, as one might expect, at the beginning, and it is entirely reasonable that they should be used to explain set theory to beginners. Here one has to consider beginners of two different types: both absolute beginners with the concept of set, who nowadays might be elementary school children, and beginners in *axiomatic* set theory, who have some familiarity with the set concept and its use in mathematics, but who are beginners in set theory as a systematic axiomatic theory. It is the latter audience that is addressed in the type of exposition commented on below.

I won't try to argue here that none of these conceptions by itself is adequate to set theory, even if they are taken to be clear enough by themselves.[10] I have made large parts of the case in previous writings.[11] It is relevant to our present concern that a more structuralist conception of set theory also played a role in its earlier history in the work of Zermelo.[12] All this is a pretty good indication that an ontological conception of set theory can't be right. But it doesn't prove it. For all three of our rough and ready notions may be just crude forms of a more sophisticated and adequate notion of set, which is still richer than the structuralist view allows. In order to come to some sort of conclusion as to whether such a conception is needed, I shall examine some considerations offered in justification of axioms of set theory. These arguments, however, leave the question how they themselves should be taken. Where something more than a structuralist conception would allow is involved, is it intended literally? Does it possibly belong to a genetic account, something that belongs to the conception of set at a certain point in its development but not to that of mature set theory? And if the answer to this question is affirmative, where does that leave the matter of commitment to sets in some richer sense?

II

A common method of explaining the axioms of Zermelo–Fraenkel set theory (ZF) relies on what was called above the notion of collection, more specifically on the idea that sets are 'formed' from their elements, one could almost say 'constructed,' in stages. That this 'formation' proceeds in stages is a consequence of the ontological priority of the elements to the set. It belongs to the picture that the stages are well-ordered. Thus, speaking of Russell's paradox, Joseph Shoenfield writes:

> The explanation is not really difficult. When we are forming a set *z* by choosing its members, we do not yet have the object *z*, and hence cannot use it as a member of *z*. . . . Putting the matter in a positive way, a set *z* can have as members only those sets which are formed before *z*. . . . Carrying the analysis a bit further, we arrive at the following: Sets are formed in *stages*. For each stage *S*, there are certain stages which are *before S*. At each stage *S*, each collection consisting of sets formed at stages before *S* is formed at stages before *S*.[13]

Shoenfield proceeds to argue for the axioms of ZF using these ideas. Similar presentations can be found in other writings on the elements of set theory. These explanations make use of somewhat pictorial ideas, but the authors generally do not make clear to what extent they are to be taken literally. Thus in the explanation quoted Shoenfield talks of the 'forming' of a set from its elements by *us*. It seems that this could be a practical possibility at most in the case of finite sets, and not too large ones at that. By analogy with other parts of the foundations of mathematics, one could claim that it is possible 'in principle' to form any finite set of given objects. But already in the case of an infinite set defined by a clearly understood predicate, this way of speaking is metaphorical, at least if the set is to be formed from its elements. When we come to sets of high rank, moreover, it is very difficult to take seriously the idea that all the intermediate sets that arise in the construction of this set from individuals can be formed by us. This 'forming' would have to take place in time and could require many more stages than there are points in time. The idea that for any ordinal number time could have had a structure so rich that that ordinal could have been embedded into it is hardly plausible enough to lend intelligibility and plausibility to the axioms of set theory.

These remarks summarize a critical discussion I gave some years ago, where I also attempted to reformulate some of these justifications in such a way that this sort of objection will not apply.[14] In considering the relevance of 'intuitive' justifications of the axioms to structuralism, however, we should allow for the possibility that no such explanations will be successful and that any attempt to justify the axioms directly will make use of metaphors or other notions that cannot be taken literally. Another relevant point is that the arguments that have actually been offered have rested on different ideas that don't clearly blend into a whole. Thus some writers, with strong considerations in their favor, have seen these arguments in something like the way I saw conceptions of what a set is, that is as heuristic principles that have a certain suggestive role but which are not adequate for the justification of the whole system of axioms.[15]

On these grounds, one might question the objectivity of set theory or at least whether our concept of set is sharp enough to determine objectively all questions of set theory or to give rise to a structure determined up to isomorphism. The latter position would fit well with a structuralist conception, since it would imply that the ontological conceptions concerning sets that cause difficulties for the structuralist view are part of an explanation by analogies and not necessarily part of the literal truth about sets. Conversely, looking at the structuralist view itself, one might ask whether it concedes to set theory the degree of objectivity that many set theorists are themselves inclined to claim, following the example of Gödel in "What Is Cantor's Continuum Problem?"[16] I am not undertaking here to defend a definite view on the still actively disputed question how far the objectivity of set theory extends. I do wish to argue that to the extent that they are persuasive, the considerations offered in favor of axioms of set theory are compatible with a structuralist view.

III

At this point one might observe that the set theorist has more argument for the axioms that he uses than the direct, intuitive, and perhaps metaphor-infiltrated arguments that occur in expositions like Shoenfield's. Many of the analogies used are of a much more mathematical kind, especially in the case of large cardinal axioms where it is a matter of generalizing properties of ω and other smaller cardinals. Another very important type of argument is a posteriori: It is argued that an axiom is likely to be true because it leads to a theory in some domain that unifies a number of phenomena in a satisfying way and settles questions that cannot be settled otherwise. For example, such arguments are used for determinacy assumptions, which made it possible to extend through the whole hierarchy of projective sets the theory developed at the lowest levels of the hierarchy by the classical descriptive set theorists. An alternative in this case might have been the axiom of constructibility, which had also been shown to yield a theory, but the theory yielded by determinacy was regarded as more satisfactory and cohered with large cardinal axioms, inconsistent with constructibility, that set theorists were disposed to accept. A kind of holistic justification is possible in set theory, in which there is a certain amount of mutual reinforcement of axioms and their consequences. In this case there is also reinforcement of axioms of different kinds: determinacy axioms and large cardinal axioms.[17]

A posteriori arguments of this kind seem to me to have no definite bearing one way or the other on the question whether a structuralist view of sets is correct. The reason is simple: They undertake to justify the axioms by their logical consequences, and what these consequences are is independent of the issue between the structuralist and the ontological view of sets. It follows that one important structuralist, W. V. Quine, is immune to the objection I am considering. In his view, except perhaps for the 'obvious but false' idea that is embodied in the universal comprehension schema, the justification of axioms of set theory is entirely a posteriori.[18]

For this reason, although a posteriori arguments are of great importance for the larger question of the objectivity of set theory, I will largely leave them out of account in what follows. Also, the difficulties for eliminative structuralism already arise in application to ZF, and it codifies modes of reasoning in set theory that in their main lines had developed before the specific axioms were formulated. Moreover, some expositions seem to rest on the presupposition that ZF can be justified without a posteriori arguments; that makes it a better testing ground for the significance of more intuitive justifications. I shall therefore consider only the axioms of ZF and not large cardinal, determinacy, or other axioms.[19]

We should begin with the idea that Shoenfield expresses with his talk of formation but which at least can be divorced from the idea that sets come into being by human *action* (or perhaps by Husserlian 'acts'). A somewhat vague principle used by Shoenfield and others in explaining axioms of set theory is that a set can be formed from *any* given objects. A way of putting it is to say that any given objects can constitute a set. The latter formulation removes the implication of formation as an action. It still contains the unexplained term 'given', which is meant to express the priority of the elements of a set to the set. But in the modal formulation this term may be superfluous, because the world 'can' already expresses the fact that the set is potential relative to its elements. In other words, 'given' means roughly actual, and the principle says that there can be a set with these objects as elements. Russell's paradox will then show that in some cases there is not an *actual* such set.

The modality here would still give rise to a number of problems, but I would like now to point to another: The principle uses a plural quantifier 'any objects.' This is the point where Cantor used the terms *Vieles* and *Vielheit* and other writers have used such terms as 'multiplicity' and 'multitude.'[20] In our earlier terms, the notion of plurality is invoked. If, with George Boolos, one regards such plural quantification as clear and ontologically noncommittal even in such an abstract setting as this, one will not see a problem.[21] I have argued that it is at least not ontologically noncommittal,[22] however, and other views of the semantics of such plural quantification in natural languages introduce reference to sets or classes. In particular cases, however, either the plural quantifier can be cashed in by first-order quantifiers or our principle about sets takes the form of a comprehension principle. A case of the former kind is that of the axiom of pairing. Given objects x and y, our principle says that they can constitute a set, that is, a set whose elements are just x and y.

It should be noted that if we formalize the preceding instance by a modal quantificational logic of the usual kind, the variables will have objects assigned to them rigidly, so that (viewing the matter in terms of possible-worlds semantics) in a possible world in which the set $\{x, y\}$ exists, its elements will be the very x and y for which it is claimed that a pair set can exist.

In the general case, however, Boolos's way of reading second-order formulas is helpful. Our principle could be stated a little more explicitly as follows: For any objects there can be a set such that every object is an element of it if and only if it is one of them. This suggests something that is essential if the formulation is to escape Russell's paradox: The plural quantifier is outside the scope of the modal operator; what objects are referred to is settled before one considers alternate possibilities. If one thinks of the formula in the usual syntax of second-order logic – that is, as

$$\forall F \ \Diamond \exists y \ \forall x \ (x \in y \leftrightarrow Fx),$$

then with the interpretation of second-order variables usual in second-order modal logic one has to either have scope operators that put 'Fx' outside the scope of '$\Diamond$' or, for arbitrary F, assume only that the possibility claim holds for an equivalent of F that is fully rigid.[23] The intuitive conception of pluralities or 'multiplicities,' however, we have taken to imply full rigidity.[24]

Let us return to the very simple case of the axiom of pairing. Does our relating it to the general principle tell us anything about what the set is that either confirms or disconfirms the structuralist view? The principle surely is not informative by itself, without either explicating its content further or examining why it seems evident. One way of reading the term 'constitute' is so that in the case at hand x and y are constituents, that is to say analogous to *parts*, of the set $\{x, y\}$. This way of thinking about the set–element relation can be found in Cantor and, given the prominence of conceptions about wholes and parts in both ordinary thinking and in the metaphysical tradition, derives no doubt from the origins of the mathematical concept of set.[25] If it is regarded as a general mark of the element–set relation, it motivates the priority of element to set that is the keystone for the accepted strategy of avoiding the paradoxes.

Although this analogy of the set–element and whole–part relations may contribute to our understanding of the set-theoretic notions and even therefore to the evidence of principles such as the pairing axiom, the question is really whether it is essential that sets be thought of in that way. I would argue that it is not. For just the pairing axiom, this is even obvious, since it is certainly true on the conception of 'sets' as extensions of predicates. But there are other intuitive representations of the set–element relation that do not rely on the whole–part concept. One frequently appealed to by set theorists is the conception of sets as trees, where the set is the root, its elements are immediately connected with it, their elements are immediately connected with them, and so on, so that the end points either represent the empty set or are 'labeled' by individuals. To make this intuitive, one is likely to think of it spatially, although that brings in some extra structure (the typical plane representation in diagrams of trees would entail an ordering of the elements of a set) and, what is more important, it does not generalize to higher cardinalities. The intention behind this kind of representation seems to me really structuralist: It illustrates the idea that what matters among sets is which ones are connected with which others by the set–element relation. It seems to me that one might also think of the 'formation' of sets from given objects as a kind of labeling of the objects, such that to each object (many) labels are attached with a proviso of extensionality: No two distinct labels can agree with respect to what objects they are attached to. This picture also has a kind of structuralist character: The principle 'Any objects can constitute a set' is taken to mean that objects ('labels') can be assigned to given objects in a many-to-one way, quite arbitrarily except for the proviso of extensionality. Thus it seems to me that quite generally, the principle that any given objects can constitute a set does not depend for its persuasiveness on a conception of what sets are that is incompatible with structuralism; rather, the existence of different ways of expressing it 'intuitively' tend rather to favor the structuralist conception.

One can also see the relation between the set–element relation and the whole–part relation in a somewhat different way, possibly more satisfactory because in terms of it the connection of set theory with a formal theory of the whole–part relation, i.e., mereology, has been worked out by David Lewis.[26] According to this conception the parts of a set are its nonempty subsets, so that one-element sets are mereological atoms, lacking proper parts. In a monadic second-order version of mereology, it is possible to develop set theory with a single set-theoretic primitive, the singleton relation $y = \{x\}$. On this account, the analogy to the whole–part relation of the relation of $\{x, y\}$ to x and y is condemned as misleading: $\{x\}$ and $\{y\}$ are literally parts of $\{x, y\}$, and x and y themselves can seem to be so only by confusing an object with its singleton. In particular, x is in no way a part of $\{x\}$. One could argue for this by the observation that if x were a part of $\{x\}$, it would have to be a proper part, since otherwise the distinction between x and $\{x\}$ would collapse and we would not have set theory but just mereology; but it is hard to see how x could be a proper part of $\{x\}$.

In his book Lewis explains his conception of set theory in a thoroughly ontological way. One could not cite him as an authority in favor of structuralism. It is noteworthy, however, that the ontological conception from which he starts is quite different from those appealed to in the earlier history of the subject. The kind of constitution by elements peculiar to the conception of sets as collections could be assumed only for one-element sets, and that must, for reasons just given, be distinguished from the part–whole relation.[27] The rest of the work is done by mereology. That is, however, a somewhat misleading statement because the mereology Lewis uses is a monadic second-order theory. The second-order quantifiers are explained as 'plural quantifiers'; in our terms, Lewis appeals to the notion of plurality. But a plurality is not a set. To assume that any objects can constitute a set is not only not part of Lewis's theory but would not be easy to reconcile with it.[28] A lot more should be said about Lewis's treatment of the foundations of set theory. But provisionally, we can find in it some support for our view, since he arrives at ZF on the basis of ontological intuitions that differ importantly from those usually appealed to. It should be pointed out, however, that Lewis does make use of the idea of 'limitation of size', of which more shortly.[29]

IV

I shall now consider two cases that are harder for the justification of the axioms whatever one's point of view on the issue that immediately concerns us. These are the axiom of replacement and the power set axiom. In both cases, there are intuitive plausibility arguments for these axioms, but arguments of different kinds are in the literature. There is not complete agreement about the 'intuitive basis' of either, and that inclines to the conclusion that a posteriori considerations play an important role in the acceptance of both.

Informally, the axiom of replacement says that if we have a set a and a relation that to an element x of a associates at most one object R_x, then there is a set whose elements are the R_x. A picture is that we can in a 'replace' each element x by its counterpart R_x (deleting it if it has no counterpart), and the result will be a set. The axiom of replacement was proposed independently by Fraenkel and Skolem at about the same time. The precise formulation used today is due to Skolem, but this more intuitive conception of the axiom is due to Fraenkel.

In the iterative conception, the objects R_x will be objects that are available at certain stages; this will be the case if they are either individuals (assumed to be available prior to the 'formation' of any sets) or sets.[30]

One type of argument for the axiom proceeds by considering, for each $x \in a$, the stage S_x at which R_x is 'formed' (more literally, the first stage at which all the elements of R_x are available). Shoenfield, for example, claims there will be a stage S after all the S_x for $x \in a$.[31] Clearly all the R_x are available before S; hence a set consisting of all of them can be formed at S. Unfortunately Shoenfield does not argue for the claim that there will be such a stage S; in his book *Mathematical Logic* he states the principle involved as a "confinality principle" and argues that since the set a can be "visualized as a single object," the same is true of the stages S_x for all $x \in a$.[32] But that visualizability as a single object is preserved by replacement of x by S_x is, given the point of view of Shoenfield in this work, just the principle of replacement itself.[33]

This argument might be interpreted in two ways: The claim that the S_x can be visualized as a single object may be taken to imply that there is a set of all of them, and then S is just the stage at which it is formed. Then Shoenfield is literally assuming a case of the principle of replacement. The other reading, however, could be simply to take the assumption as one about giveness or 'availability' of objects: The elements of a are all 'given'; and this possibility of being simultaneously available is what is preserved by 'replacement'. Then S is just the stage at which all the S_x are available. If we look at the argument this way, then we can proceed (as does Hao Wang[34]) without reference to stages: Since the elements of a are given objects, so also are the R_x, and therefore they can constitute a set. One way of thinking of the axiom of replacement suggested by these considerations is that what underlies it is a principle about when objects can be considered as given or 'available' or, in Cantor's language, existing together.[35] Then although an ontological intuition of a kind is being appealed to, it is not one specifically about *sets* and thus not incompatible with a structuralist conception of sets.

The axiom of replacement arises naturally from the idea of 'limitation of size,' which has often been considered independent of the iterative conception. A crude way in which the idea is expressed is that a collection is a set if it is not too large, and that a collection fails to be a set if it *is* too large. This formulation uses the term 'collection' in a generic way (cf. note 9), in particular, obviously, so as not to presuppose that collections are sets. But clearly other notions such as class or plurality can be substituted if one attributes to them the requisite generality. Applications of the idea are best understood if we think of it as a second-order principle, in the general sense that it involves generalization of predicate places in our language, whether or not we take the further step of formulating it in terms of second-order logic.

At least if we think of 'too large' in terms of cardinality, the axiom of replacement falls rather directly out of the limitation-of-size idea, since if a set a is given, then *it* is not too large, and there are not more R_x's for x in a, so that the 'collection' of them cannot be too large either. (The second-order reasoning used here has minimal presuppositions.[36] A more serious question concerning replacement is what counts as an admissible relation; the standard first-order version is already quite strong, since it admits nesting of quantification over the whole universe of sets.)

Some care needs to be observed with the limitation-of-size idea. It has in the past been understood as a kind of device for separating sheep from goats among collections or pluralities, or as a criterion for a predicate to have an extension, either way arising from a diagnosis of the paradoxes. This also makes it more specific to sets than seems to me warranted. Behind this understanding of the principle lies an interpretation that represents the universe of sets as a determinate reality and quantification over it as not different from other forms of quantification.

On the contrary, 'limitation of size' is more properly understood in terms of the idea of inexhaustibility, emphasized by Gödel as early as 1933[37] but in a way implicit in Cantor's distinction between the transfinite and the absolutely infinite. In a sense, there are no 'collections' that are 'too large' to be sets. If we try to form an idea of such a collection, as soon as it assumes a certain degree of definiteness we become able to conceive of a still larger collection. For example, we try to describe the universe of sets by its containing the natural numbers or some other basic infinite set and being closed under the operations of power set and replacement (taking for granted more elementary operations). But it is arbitrary to suppose that this gives us everything, we have already since 1930[38] a clear conception of a *set* closed under those operations: the set of sets of rank less than α, for α a strongly inaccessible cardinal.

This way of looking at 'limitation of size' illuminates the role of the power set axiom, which as we shall see is difficult to justify intuitively. We can't show, starting with the limitation of size conception and the other ideas I have canvased, that the sets of integers would *not* be a too large totality or that they are not inexhaustible as the ordinals or as the sets themselves are. As Michael Dummett observes, the concept 'set of natural numbers' could be an indefinitely extensible concept, and the different forms of predicativism present a perfectly coherent conception according to which that is so. But the power set axiom does *cohere* with the limitation of size conception in a satisfying way, because it provides a way of going to larger and larger sets. Without it, we could not prove it false that all sets are either finite or denumerably infinite.

The idea of limitation of size does not make any demand on the concept of set that is incompatible with structuralism. We might note that the inexhaustibility that underlies it applies as well to ordinals and cardinals as to sets. But the main point is that it is a maxim that tells us something about when sets exist, without requiring anything further of what they are to be, that is, that there will be a set *a* such that anything is an element of it if and only if a certain condition holds, and no more. One could thus view it as a principle about extensions (the most 'structuralist' of the three conceptions mentioned in Section I), but in its application it is combined with principles that the conception of extension would not give rise to. If we think in genetic terms, limitation of size tells us that we can go on, that there is more to the 'universe of sets' than we have so far got hold of, but it doesn't tell us that this 'more' is more than a richer structure, more objects some of which stand in the membership relation to some others.

I now turn to the power set axiom, which is probably the most difficult case for the whole enterprise of justification of axioms of set theory by intuitive considerations. There is no agreed upon approach. The power set axiom does not arise directly from an intuitive conception in the way in which, for example, Pairing arises from the idea that any objects can constitute a set, or Replacement from the idea of limitation of size. It is, to be sure, a pillar of the iterative conception. But in practice that conception *assumes* the power set operation; according to it, sets are what is obtained by iterating the power set operation starting with an initial domain of individuals.

It is, of course, possible to understand the iterative conception in a weaker way, according to which sets are obtained by iterating the operation of *forming arbitrary subsets*. Then, however, suppose we have at a certain stage *S* a set *a*. Then any subset of *a* can be formed at the next stage, but that does not tell us that the formation of subsets of *a* will not stretch out indefinitely through the stages, so that there is never a stage at which all the subsets of *a* are available, which would be necessary to form the power set. In other words, this version of the conception does not rule out the idea that the totality of subsets of a set is irremediably potential, like the totality of sets itself.

At this point writers on the subject appeal to a principle of plenitude: If *a* is formed at a certain stage *S*, then its elements are available at *S*, and therefore any subset of *a could* have been formed at *S*. It is then concluded that all subsets of *a are* formed at *S* and are thus available at the next stage, at which $\mathfrak{P}(a)$ can be formed.[39]

This principle makes a nice connection between the power set axiom and an idea from the metaphysical tradition, but it is not convincing in itself. Why should we accept the appeal to plenitude in the case of subsets of a set and not in the case of sets in general or ordinals?

The situation is not essentially changed by Paul Bernays's classic description of reasoning about arbitrary sets and functions of natural numbers as 'quasi-combinatorial': By analogy of the infinite to the finite, we "imagine functions engendered by an infinity of independent determinations which assign to each integer an integer, and we reason about the totality of these functions."[40] As Bernays implicity recognizes, the analogy in question has two distinguishable aspects: an understanding of an arbitrary function on the natural numbers (or of an arbitrary set), and the further step of regarding the "totality of these functions" as completed, or as giving rise to a set. The weak iterative conception can appeal to the first aspect of the analogy, and then the question raised concerns the second. Both for their argumentative force and for their relevance to the structuralist view what is important is that what is offered are *analogies*, so that not everything in the picture is intended to be literally true. In particular, we cannot take Bernays's "independent determinations" literally as *choices* (which would certainly be incompatible with structuralism).

I think we have to admit that an essential part of the justification of the power set axiom is a posteriori. It enables us to formulate in terms of set theory the ideas about the continuum that underlie and come out of classical analysis, in particular as it was reformulated in rigorous form in the nineteenth century.[41] Once one accepts the continuum as a set, the power set axiom is a general principle to which the set-theoretic construction of the continuum can be reduced. Power sets of denumerable sets already present the conceptual problems that the idea of power set gives rise to. It would be arbitrary to accept the reals or the set of sets of natural numbers as a set and not to accept the power set operation in general. And as we have noted above, acceptance of the power set coheres with other principles motivating the axioms of set theory.

This discussion of the arguments that are actually in the literature should make plausible that there is not a set of persuasive direct 'intuitive' considerations in favor of the axioms of ZF that are incompatible with a structuralist conception of what talk of sets is. If we add to this the neutrality of a posteriori considerations with respect to this issue, we can conclude that an ontologically richer conception of set is not needed for ZF. Although I have not argued this here, in my view the considerations that are offered in favor of axioms going beyond ZF are not different in this respect. But it should be clear that this paper has undertaken to answer a specific objection to the structuralist view of mathematical objects and not to mount a general defense of it.

NOTES

1. "Mathematics as a Science of Patterns: Ontology and Reference." *Noûs* 15(1981: 529–50, p. 530.
2. A classic instance of set-theoretic structuralism is Dedekind's account of the natural numbers in *Was sind und was sollen die Zahlen?* (Braunschweig: Vieweg, 1888); an example of the effort to avoid the concept of set in stating a structuralist view is Resnik, "Mathematics as a Science of Patterns." What has been variously called "if-thenism," "implicationism," or "deductivism" would be an example of non–set-theoretic structuralism, but in the last twenty years this approach has hardly found any defenders.
3. "The Structuralist View of Mathematical Objects," *Synthese* 84 (1990): 303–46.

   The ambiguity mentioned in the text is related to but distinct from another problem in recent discussions of structuralism in the literature, whether structuralism is an alternative to, or a form of, platonism. The view of "The Structuralist View," after what I called "eliminative structuralism" has been rejected, is that although structuralism is intended to avoid certain pictures associated with platonism, it is in the most fundamental sense platonistic, at least in application to strong theories such as set theory. In this I am in agreement with Resnik.
4. "The structuralist View," §7. See also *Mathematics in Philosophy* (Ithaca, N.Y.: Cornell University Press, 1983), pp. 22, 191–2, where, however, I did not distinguish between a strictly structuralist conception of the universe of sets and the 'ontological' view to be stated presently, where the latter view allows for ontological relativity and 'incompleteness'.
5. This would, to be sure, offer the consolation that we could save eliminative structuralism in application to other kinds of objects (such as those of pure geometry) by denying it in the case of sets. Of course we would not thereby save eliminative structuralism as a program for eliminating all mathematical objects or even all pure mathematical objects.
6. The principle that there is a set of all individuals requires separate discussion, since it implies that there are not as many individuals as there are sets. On the point of view just stated, however, it would be true for most realizations that would be all natural.

   If we think in terms of the formation of sets from their elements, then it is natural to think of the individuals as 'available' prior to the formation of any sets. Then that there is a set of them is a consequence of the principle that any available objects can constitute a set. On the usual understanding of an individual as an object other than the null set that has no elements, it might be thought a mathematical accident that this point of view is workable: Other objects that must be posterior to some sets, such as ordinals and functions, can as it turns out be represented as sets. However, if this were not so the remedy might be just to revise the definition of individual.

7. An objection along these lines was sketched to me very briefly in conversation several years ago by Timothy McCarthy. Whether its further development here corresponds to what he had in mind I do not know.
8. See part III of "Genetic Explanation in *The Roots of Reference*," in Robert Barrett and Roger Gibson (eds.), *Perspectives on Quine* (Oxford: Blackwell, 1990), pp. 271–90.
9. What I, following Frege, call extensions is what some writers call logical collections, in effect using the term 'collection' as a generic term for any entity that might do duty for sets or classes. An example is Penelope Maddy, *Realism in Mathematics* (Oxford: Clarendon Press, 1990), pp. 102–3. That usage seems to leave us without a term for what I call a collection, at least if one distinguishes, as I argue in the text should be done, between the notion of collection and that of plurality. Maddy's term 'combinatorial collection' (ibid., p. 102) is close to my term 'collection', but her usage is more specifically tied to the iterative conception of the universe of sets.
10. The latter has been questioned especially for the notion of collection. Frege's criticisms of some of his contemporaries amounted to the claim that this notion could not be clearly distinguished from that of a mereological sum. Nelson Goodman's well-known rejection of classes is again in the first instance of rejection of collections. (Cf. "Genetic Explanation," p. 284.) See also Max Black, "The Elusiveness of Sets," *Review of Metaphysics* 24 (1970–1), 614–36. In my terms, Black is severe with the idea of collection but friendly to that of plurality.
11. In particular "Sets and Classes" and "What Is the Iterative Conception of Set?" (Essays 8 and 10 of *Mathematics in Philosophy*) and "Genetic Explanation."
12. "Genetic Explanation," pp. 285–6.
13. "Axioms of Set Theory." p. 323. In a footnote Shoenfield states that 'before' is to be understood "in a logical rather than a temporal sense."

    It is possible, given other assumptions about stages, to derive the statement that they are well ordered (which will yield the axiom of foundation) from the assumption that they are partially ordered. See "Axioms of Set Theory," p. 327. (This was first shown by Dana Scott.) For fuller discussion see George Boolos, "Iteration Again," *Philosophical Topics* 17 (1989): 5–21. I doubt that this somewhat remarkable technical fact is of much philosophical significance, since it is hard to see how the existence of a partially ordered sequence of stages is more evident than the existence of a well-ordered sequence.
14. "What Is the Iterative Conception of Set?"
15. I take this to be the intention of the somewhat skeptical discussion of such justifications by Maddy, "Believing the Axioms," *The Journal of Symbolic Logic* 53 (1988): 481–511, 736–64. Similarly Boolos, "Iteration Again," sees some axioms as coming naturally from the 'iterative conception of set' and others quite independently from the idea of limitation of size.
16. In *Collected Works, Volume II: Publications 1938–1974*, ed. Solomon Feferman et al. (Oxford: in Oxford University Press, 1990). The article mentioned in the text was first published in 1947, and revised and expanded in 1964.
17. This, however, has turned into a kind of reduction, since it was shown by Martin, Steel, and Woodin that a strong large cardinal axiom, the existence of a supercompact cardinal, implies that determinacy holds in $L[R]$ (the sets constructible

relative to arbitrary sets of integers); this is the strongest determinacy assumption used in work in descriptive set theory. I don't think, however, that this reduction abolishes the mutual reinforcement of determinacy assumptions and large cardinal axioms, since that it yields given determinacy principles and their consequences in descriptive set theory would be offered as an a posteriori argument for a large cardinal axiom. On the variety of considerations used by set theorists in justifying their assumptions, see Maddy, "Believing the Axioms."

18. This view is the one most explicitly expressed by Quine and the dominant view before *The Roots of Reference* (La Salle, Ill.: Open Court, 1974). The latter work makes some concessions to more usual ways of looking at set theory. See my "Genetic Explanation."
19. Although the axiom of choice is not considered below in spite of the fact that it is treated by most set theorists as part of basic set theory and thus comparable to (the rest of) ZF, I do not believe it would raise any new problems.
20. Cantor, *Gesammelte Abhandungen*, ed. Ernst Zermelo (Berlin: Springer, 1932), pp. 204, 443 and elsewhere; Wang; *From Mathematics to Philosophy* (London: Routledge, 1974), pp. 281–4. The choice of language already seems to force a choice as to whether one will regiment the plural by the singular. The more conservative choice is not to do so, and I will try to avoid it.
21. Boolos, "To Be Is to Be the Value of a Variable (or to Be Some Values of Some Variable)," *Journal of Philosophy* 81 (1984): 430–49, and "Nominalist Platonism," *Philosophical Review* 94 (1985): 327–44. See also David Lewis, *Parts of Classes* (Oxford: Blackwell, 1991), §3.2.
22. "The Structuralist View," §6.
23. *Mathematics in Philosophy*, pp. 315–18.
24. Thus if we assume that the second-order variables range only over pluralities, we can use (1) without modification. Moreover, we can simplify the second-order logic. We would have for the second-order variables rigidity *axioms* (the RS, R$, and BFS of *Mathematics in Philosophy*, p. 335), and the comprehension schema would be expressed just as in nonmodal second-order logic, i.e., as follows:

$$\exists F \forall x_1 \ldots x_n[Fx_1 \ldots x_n \leftrightarrow A],$$

(A is any formula; typically, but not necessarily, $x_1 \ldots x_n$ are just the free variables of A). But its force is not that of the usual comprehension axiom in second-order modal logic, but rather of extensional comprehension [ibid., p. 336, from Gallin, *Intensional and Higher-Order Modal Logic* (Amsterdam: North-Holland, 1975), p. 77]. This simplification, unfortunately, is of limited use for the axiomatization of set theory in a modal language. To obtain the force of the usual axioms of separation and replacement, we need to instantiate for the second-order variables predicates that are not fully rigid. That situation is likely to obtain more widely than for the particular theories developed in Essay 11 of *Mathematics in Philosophy*.

25. One can see this clearly in one of the earliest explanations of the concept of set as a mathematical concept, in Bernard Bolzano's *Wissenschaftslehre*, §82; see also *Paradoxien des Unendlichen*, §3.
26. Lewis, *Parts of Classes*; see also "Mathematics Is Megethology," *Philosophia Mathematica* (Series 3) 1 (1993): 3–23.
27. Lewis seems to me in fact not to rely on the idea of the priority of the elements of a set to the set. The paradoxes are avoided by two features of the theory: the

fact that comprehension occurs in the setting of second-order logic (so that it does not yield a Fregean extension) and the use of the limitation of size idea, which is ingeniously adapted to the mereological setting.

Lewis does consider a view he calls structuralism; it is in effect eliminative structuralism applied to the specific case of the singleton relation. Because he allows himself only monadic second-order logic as a framework, this poses a technical problem of how to obtain dyadic second-order quantification. Solutions (dependent on mereological assumptions about how close 'Reality' is to being atomic) are offered in the Appendix to *Parts of Classes* by Burgess, Hazen, and Lewis.

Should Lewis's construction lead me to reconsider my rejection of eliminative structuralism as an approach to set theory? Even if one grants him monadic second-order mereology as in effect a part of logic, and the atomicity assumptions needed to obtain dyadic second-order quantification, the answer is no. For the question still arises what domain of objects could possibly make out the possibility of the structure satisfying the formal properties of the singleton relation with all the mereological fusions the logic than yields. Lewis's own answer to this problem is in effect that such an hypothesis is needed for standard mathematics. I can accept this answer, but then the distinction between an eliminative and noneliminative structuralism about sets as Lewis conceives them virtually disappears.

28. Our own reading of that principle would fit uneasily into Lewis's metaphysics, because of his reduction of modality to possible worlds.
29. In "Mathematics Is Megethology," Lewis takes a more structuralist view.
30. Although this does not happen in the usual way of doing set theory, it would be theoretically possible that there could be objects other than sets that are generated in stages; for example, to avoid the Burali–Forti paradox it is natural to think of ordinals that way quite independently of the von Neumann treatment of them as sets. Cf. note 6.
31. "Axioms of Set Theory," p. 326.
32. Reading, Mass.: Addison-Wesley, 1967, p. 240.
33. As already argued in *Mathematics in Philosophy*, p. 280, n. 16.
34. *From Mathematics to Philosophy*, p. 186; cf. *Mathematics in Philosophy*, pp. 279–80.
35. In one kind of modal theory, the "theories of potential sets" of *Mathematics in Philosophy*, the role of the axiom of replacement is taken by a purely modal-logical reflection principle.
36. In a second-order language, we can state the principle of limitation of size as an axiom: A plurality is a set if and only if it is not equinumerous with the universe. This axiom implies the axiom of replacement (and of course separation). A more anomalous consequence is that it implies that the universe is well-ordered: For by the Burali–Forti paradox, the ordinals are not a set; hence they must be equinumerous with the universe. That is, there is a one-to-one map of the ordinals onto the whole universe, which induces a well-ordering of the latter.

    The axiom in question was put forth (in a variant formulation) as Axiom IV.2 in §3 of J. von Neumann, "Eine Axiomatisierung der Mengenlehre," *Journal für die reine und angewandte Mathematik* 154 (1925): 219–40; cf. the discussion in §2.
37. "The present situation in the foundations of mathematics," lecture given to the Mathematical Association of America, forthcoming in *Collected Works, vol. III: Unpublished Essays and Lectures*, ed. Solomon Feferman et al. (Oxford: Oxford University Press).

38. Cf. Ernst Zermelo, ''Über Grenzzahlen und Mengenbereiche,'' *Fundamenta Mathematicae* 16 (1930): 29–47.
39. Such an appeal to plenitude is implicit in Shoenfield, ''Axioms of Set Theory,'' p. 326. It is embodied in axiom IX of George Boolos, ''The Iterative Conception of Set'' (first published in 1971), in Paul Benacerraf and Hilary Putnam (eds.), *Philosophy of Mathematics: Selected Readings*, 2nd ed. (Cambridge: Cambridge University Press, 1983), p. 494.
40. ''Sur le platonisme dans les mathématiques,'' *L'enseignement mathématique* 34 (1935): 52–69, p. 260 of the translation in Benacerraf and Putnam, *Philosophy of Mathematics.*
41. Shaughan Lavine argues persuasively that the power set was not assumed in the theory of sets and transfinite numbers as Cantor originally worked it out and that reasoning that would require the power set axiom appears in his work only after 1890. In Lavine's view, Cantor's assuming at that point of the equivalent of the power set axiom led to the consequence that the real numbers are a set, something that Cantor had earlier hoped to prove by, in effect, proving the Continuum Hypothesis. See *Understanding the Infinite* (Cambridge, Mass.: Harvard University Press, forthcoming), chap. IV.

7

The Kant–Frege–Russell View of Existence: Toward the Rehabilitation of the Second-Level View

DAVID WIGGINS

By whatever and by however many predicates we may think a thing – even if we completely determine it – we do not make the least addition to the thing when we declare that this thing is.

Kant, *Critique of Pure Reason*, B628

The best atheist joke: "God's only excuse is that he doesn't exist."

Nietzsche, *Ecce Homo*,
Penguin 1971, p. 58 (trans. Hollingdale)

1. Ruth Barcan Marcus befriended certain true doctrines at times when their prospects of general acceptance, or even of visibility, seemed vanishingly small. By scorning the scorn of those who thought all truths ought to be turned out from a mold they found familiar, and by making ideas fully visible that few at the time were disposed even to consider, she influenced decisively (however belatedly, given the early date of some of her important papers) the development of logic and philosophical logic in the postwar period. Through that development something distinctive came from modern logic and its modal extensions (just as she had envisaged that it would) into traditional metaphysics.

The moral for the rest of us to draw from this history of vindication is not so much that we should always expect what is not to be expected (it is too easy to repeat and too difficult ever to follow that old Heraclitean saw) but rather perhaps this: That where the rival of a current doctrine is almost visible, we should make it visible and try to secure its equal development. This is not to say that we must insist on the rights to audience of intellectual idiocy. If the rival to an accepted view turns to idiocy even as you attempt its equal development, abandon it forthwith.

2. At the acme of the influence in philosophical logic of Russell and Frege, few can have predicted that, after their deaths, the idea that they both sought to discredit of existence as a first-level concept (or property of individuals) would so soon be restored. Nor can they have imagined that the methodological precept I have just mentioned would come to require the rehabilitation and redeployment of their formulations of their positive view of these matters. This was that the existence of an item (or items) x of level n is always a property of x's concept, which is a level $(n + 1)$ concept, not a level $(n + 2)$ property of x itself.[1] But the situation nobody predicted would appear to be the actual one. As something believed and positively advocated in its Fregean purity, the Frege–Russell view has virtually disappeared from sight.

Such an outcome might prompt an adherent of the Frege–Russell view to try to refute the new first level doctrine. But, in the absence of any attempt to refurbish the Frege–Russell account or resolve its difficulties, that would be absurdly premature. It would be still more premature to try to explore the relation of the restated Frege–Russell account to the substitutional interpretations of quantifiers whose possibility our admired honorand has so long and persistently defended against cavil and misrepresentation. The most I shall attempt here is to make the Frege–Russell view visible again.

3. One who would refine the old view of 'exists' has the duty to begin by admitting frankly that many who take the modern view of existence as a first-level property have been encouraged in this by wildly implausible denials that one finds in Russell and Frege, to the effect that sentences like 'There is Julius Caesar' and 'Julius Caesar exists' do not make sense. Understandably, those who take the first-level view have set out from the commonsense finding that such sentences do make sense. A predicate that the modernists will then discern there and find predicated of Julius Caesar is the first-level complex predicate '[ ] has something identical with it', 'there is something identical with [ ]' or '$(\exists x)$ $(x = [\ \ ])$'.

Identifying this first-level predicate with 'exists', as it occurs in 'Julius Caesar exists', either (I) the modernists can claim that 'exists' means something slightly different in 'Julius Caesar exists' from that which it means in sentences like 'There exist tame tigers', 'There exist some (*sǝm*) tame tigers', 'Tame tigers exist', 'Some (*sǝm*) tame tigers exist', identifying only the second kind of 'exist' with the quantifier $(\exists x)$ $(\ldots x \ldots)$ that is deployed in '$(\exists x)$ (Tame x & Tiger x)''[2]; or, option (II), they can offer the following unified account of 'exists'. First, 'exist(s)' is distinguished from the ordinary quantifier 'there is . . .' or '$(\exists x)$ $(\ldots x \ldots)$'. Then the complex first-level predicate '$(\exists x)$ $(x = [\ \ ])$', which we have already encountered, is deployed in the understanding of 'Julius Caesar exists'. Or else, for this same purpose, a first-level primitive predicate having just the same extension is introduced. Either way, a first-level existence predicate, abbreviated as **E**, is then deployed again in the understanding of 'Some [*sǝm*] tame tigers exist',[3] and may just as well be applied to the cases of 'There exist tame tigers'/'There exist some (*sǝm*) tame tigers'/'Tame tigers exist', each and all of these being rendered as:

For some *x*, *x* is a tame tiger and *x* exists,

(∃*x*) (Tame tiger [*x*] and **E**[*x*])

or, in the binary quantifier notation that Evans rightly prefers,

(Some *x*) (Tame tiger [*x*], **E** [*x*]).

Let me say at once that I hear in these renderings, whose structure so strongly suggests that one ought to be able to say them aloud in English as '[*sŭm*] tame tigers exist', a strange account of the '[*səm*] tame tigers exist' with which I began. Surely this was a partitive use of 'some'. Is it not odd then to register 'exists' separately (as Gareth Evans does, for instance) by a distinct clause? Still, the truth condition that (II) delivers is not incorrect, truth-conditionally speaking. So I shall try not to make too much too soon of the objection. (Still less am I preparing to frame any simple or general theory about [*səm*] and [*sŭm*].) All I am meant to be doing at this point is listing the options.

4. Option (I), which proposes that we simply tolerate an ambiguity in 'exists', and option (II), which postulates no ambiguity but assimilates [*sŭm*] and [*səm*], are both possible views. But the maxim of equal development will now urge the redeployment and reformulation of the view (III), which (I) and (II) are in reaction against, namely Russell's and Frege's. (Like II, it postulates no ambiguity in 'exists'.)

Making the minimum departure from Frege, let us begin the explanation of option (III) like this (cp. "Concept and Object,"[4] but note that what is displayed here is only partly quotation):

> [When we consider sentences like 'there is Julius Caesar' 'Julius Caesar exists' or] 'there is only one Vienna', we must not let ourselves be deceived, because language often uses the same word now as a proper name now as a concept-word: in ['there is only one Vienna'] the numeral indicates that we have the latter. 'Vienna' is here a concept word, like 'metropolis'. Using it in this sense, we may also say 'Trieste is no Vienna' [or 'Vienna exists'].

What concept is this individual concept of Vienna? It is the concept *the same as Vienna*, or *identical with Vienna* – the concept that only the city of Vienna can fall under.

According to (III) then, there is no need to postulate any ambiguity in 'exists'. 'Exists' is the same verb and functions in the same way in 'Vienna exists', in '[*səm*] tame tigers exist' and in 'There exist tame tigers', where the 'there' produces the same overall effect as [*səm*]) would. At this level, 'exists' always means '(∃ *x*) (. . . *x* . . .)'. But singular terms (singular terms as syntactically identified) can occur in two different modes. And this last is a point that will need to be made independently plausible by collateral considerations such as those to be derived from 'There is only one Vienna,' 'Castor and Pollux are two', 'The Muses are nine', 'The Scots are numerous'.[5] Proper names in ordinary language may make mention of individual objects *or* concepts of individual objects. Context determines.

5. That then is what the friend of the Frege–Russell view can do with the concept *identical with Vienna* and the existential quantification that is expressed by '$(\exists x)(\ldots x \ldots)$'. But he can do more. He can reach out to the conviction that it is perfectly possible to say, without contradicting oneself, that there are all sorts of things that don't exist – dragons, Vulcan (the putative planet postulated by Leverrier), Scylla, Charybdis, for instance. (Can he say even that there are things that there aren't?) What it seems 'thing' does here (with an obligingness altogether characteristic of it) is to make it possible to quantify over first-level concepts, both general ones, the concept of dragon, and individual ones, the concepts of Vulcan, Scylla, Charybdis, and so forth, so that we can address the question of their instantiation or noninstantiation.[6]

6. So far so good. (III) is emerging as an option, at least, – and well short of idiocy. What else does the friend of the Frege–Russell view need to attempt if there is to be a reasonable view (III) to pit against views (I) and (II)?

Well, the friend of Frege and Russell now needs to come to terms with a difficulty there has always been in expounding option (III). If we say 'exists' is not a predicate of individuals but a second-level predicate, a predicate applicable only to what first-level predicates like 'tame tiger' (or 'identical with Caesar') stand for, then how shall we avoid the conclusion that what the sentence 'tame tigers exist' says is that the concept *tame tiger* exists? To say that this concept exists, however, or that there is such a concept, is quite different from saying that the concept is instantiated. (There exist indefinitely many uninstantiated concepts.)

A linguistic difficulty calls for a linguistic answer. The linguistic answer is that 'exists' or 'is' or 'are', as they occur in English, are not what they seem, but are short for something we might convey by saying 'There exist(s) ⟨of⟩' or 'there is (are) ⟨of⟩'. The ⟨of⟩ that I write here is simply intended to mark a distinctive mode of combination in which existential verbs can occur with words that Frege called concept-words. (But compare the French partitive *de* in 'Il y a des tigres domestiques' or the privative genitive in the Russian 'Ne bilo soldata v dome' ≈ 'There was no soldier in the house', literally, 'there was not of soldier in the house'. Or compare the English partitive 'of' that we find in 'instances of generosity' or the 'of' in 'an army of peasants'.) By this account, when we say that 'exists' is a predicate of what first-level predicates stand for, that is concepts, the predicate of *tame tiger* that we must intend is 'there exists ⟨of⟩' or 'there is/are ⟨of⟩'. There existing ⟨of⟩ the concept *tame tiger* is the same as the concept *tame tiger* being instantiated.[7]

Here let us revert for the moment to 'there are all sorts of things that don't exist'. Our account of this now allows us to say that there are all sorts of kinds that there doesn't exist ⟨of⟩. Which is the same as to say there are all sorts of concepts that are not instantiated.

7. At this point, we come to some of the difficulties of option (III), and the difficulties, in particular, of sustaining this theory for the case of sentences of the form 'Vienna exists/does not exist' or 'Julius Caesar exists/does not exist'. Among these difficulties, the problems have to be faced of specifying individual concepts that are instantiated or not instantiated, of assigning the right modality to statements of existence and of accounting correctly for the possibility of true negative singular existential statements.

8. In expounding option (III), I have assumed that we can express the instantiation or noninstantiation of concepts of individuals such as *identical with Julius Caesar* by prefixing the existential quantifier to a predicable [=*nn*], consisting of identity sign followed by singular term, thus

$$(\exists x)\ (x\ [\ = \text{Vienna}])$$
$$\sim(\exists x)\ (x\ [\ = \text{Julius Caesar}])$$

This was not Russell's way of specifying concepts of individuals. For reasons with which one comes to sympathize more and more as one struggles with these problems, Russell would have wanted to translate these predicables in the form [=*nn*] into pure descriptions. Since the time when Russell wrote, however, it has become much clearer just how hard it is to explain what it is for something to be Vienna or Caesar, just how hard it is to draft the specification of what is takes for something to be Caesar or Vienna, without making any reference to Vienna or Caesar themselves. The point has been effectively conceded by those who have been happy to go along with the Quinean surrogates 'Viennizes', 'Caesarizes' and so on. These surrogates – insofar as we know what they mean[8] – register very aptly the actual impossibility of dispensing with the singular terms 'Vienna' and 'Caesar' themselves.

A difficulty now presents itself, when we combine the acceptance of this last point with something that is related, namely the best theory of the meaning of singular terms. The best theory of such terms says that the sense of a genuine singular term – what one grasps when one grasps what it means – depends on the mode or modes in which the term presents the object that is its reference. The only way in which it can be stated or fixed what that reference is is by some determination of which object this is. Such a determination involves introducing a conception of that object, exploiting the larger body of background information that sustains that conception and then drawing upon both these bodies of information to make it determinate which object it is that is referred to by the term.[9] So much ought not to be very controversial. (It does not stand opposed to the thing Ruth Marcus wanted when she called proper names 'tags'. It coheres with it. To speak in the way I have here of conceptions is not to assimilate names to descriptions.) But of course the question that arises is how, when there is in fact no such object, anyone can make it clear in the way suggested which object some purported object is. Will there not be a problem of some sort wherever the only way in which we can make clear which concept it is whose instantiation or non-instantiation concerns us is by the use of language that is genuinely sufficient to say which object it is that this is the concept of? Such language will tend to be existentially committal.

9. Problems such as this are not best solved by reneging upon the hard won truth about the sense of a singular term. They will force themselves upon us again and again. But before they do, let us resolve to press on with our two sentences unreformulated, namely:

$(\exists x)$ $(x$ [= Vienna])
$\sim(\exists x)$ $(x$ [= Julius Caesar]).

The first sentence, option (III) will declare, is contingently true. The second, it will declare, is contingently false. Questions about modality will soon take center-stage, but the decision to persist in these versions makes it necessary to attend first to another point.

The articulation intended is indicated by the square brackets. The difficulty that it seems option (III) faces is this: that a rival articulation appears to be available of the very same forms, an articulation that the notation equally legitimizes.[10] Remove the preference for the articulation that is marked by the square brackets, a preference for which symbolic orthodoxy will find no place, and you can see

$(\exists x)$ $(x$ = Vienna)
$\sim(\exists x)$ $(x$ = Julius Caesar)

as affirming of Vienna and denying of Julius Caesar respectively the complex first level concept we have already mentioned, namely the concept that $(\exists x)$ $(x$ = [ ]) stands for. What this predicable stands for is something that it will seem all particulars are or do – something like breathing, J. L. Austin used to say, but much, much quieter. In memory of Austin let us call it 'ticking over.'' (See *Sense and Sensibilia*, p. 68 n.) There is nothing wrong with this predicate of ticking over. It is well defined and it is the predicate the modernists prefer to discern in these sentences.

When I make this concession, to the effect that ticking over is well defined, a friend of the modern view may react by saying this: since you admit that 'Vienna exists' is equivalent to '$(\exists\, x)$ $(x[=$ Vienna$])$', which is equivalent to '$(\exists\, x)$ $(x =$ [Vienna]$)$', you have to admit that after the deletion of 'Vienna' the rest of the sentence used in the second version, viz. '$(\exists\, x)$ $(x = [\quad])$', means 'exists'. So existence is the same as ticking over, after all.

To this I reply that this objection rests upon the fallacy Geach called the canceling out fallacy.[11] This is the fallacy by which, at its crudest and least deceptive, one could prove, from the overall synonymy of 'Socrates killed Socrates' and 'Socrates was killed by Socrates' (say), together with these sentences' equiformity in respect of their singular term positions, that what remains after the names are canceled in the one is semantically equivalent to what remains after the same names are canceled in the other. Therefore, one might conclude, killing is the same as being killed. This is only one of indefinitely many incredible things one would 'prove' by canceling out.

If the canceling out argument is fallacious, all that Frege and Russell need to insist is that $(\exists\, x)$ $(x = [\quad])$ should not to be confused with the second level predicate of existence. Ticking over is fine, but it is not the same sort of property as existing – or even a property of the same sort of thing.

10. So much for the two predicates that are there to be discerned and so much for the right choice of candidate to represent existence. Now for the defense of contingency. In its first form, the difficulty is this. We have said that it is contingently false that Julius Caesar does not exist, which implies

$$\Diamond \sim (\exists\, x)\ (x = \text{Caesar}).$$

So now consider the contingently false claim

$$\sim(\exists\, x)\ (x = \text{Caesar}).$$

Existential generalization permits us to go from any sentence in the form Φ a, where Φ if free for y, to a sentence in the form $(\exists\, y)$ (Φy). As a special case of this, it would permit us to go from our denial of Caesar's existence to a sentence in the form

$$(\exists\, y) \sim (\exists\, x)\ (x = y),$$

which represents a claim that there is something whose individual concept is uninstantiated. But is that not impossible? (Contrast this claim with the claim that there is something that doesn't exist, which we were happy to interpret as saying that there is a concept that is uninstantiated.)

The question then is this. How can a valid principle of inference carry us from a contingent falsehood to a necessary falsehood? If the antecedent of the corresponding conditional is possible, how can the consequent help but be possible? How are we to avoid the conclusion that $\Diamond(\exists y) \sim (\exists x)(x = y)$? Surely we do not intend to deny the principle $\Box(p \supset q) \supset (\Diamond p \supset \Diamond q)$.

In an alternative form, we have the difficulty that whatever object you take, there will always be something identical with it. Or

$$(\forall y)(\exists x)(x = y).$$

But, by universal instantiation, this implies

$$(\exists x)(x = \text{Julius Caesar}).$$

What is more, if (as is obviously the case) the first of these is a necessary truth, then so is the second. How then can we deny

$$\Box(\exists x)(x = \text{Julius Caesar})?$$

Or are we proposing to deny the principle $\Box(p \supset q) \supset (\Box p \supset \Box q)$?

11. The two difficulties are parallel. Let us focus first on the first. Answers to it abound in the literature. But it will be best to consider the matter afresh, without commitment to current or received codifications of quantified modal logic.[12] For these may not be the best codifications for acceptance by a friend of option (III). Of course, if there is no alternative to received codifications, then we shall find that, bit by bit, the friend has to reinvent them. (If he does, the received codifications will be strikingly vindicated.)

12. In order to avoid the first difficulty, it may seem that the obvious and right thing to do is to challenge the principle of existential generalization. Many have taken this course, and some have pursued the logos they were following all the way to the adoption of a free logic. But, if we were right to say that a name has its sense by standing for an object, that that sense is explained by saying which object the object is and that there is no saying which object this is in the absence of such an object, then it seems that, when we are *given* the truth of "Fa" together with all the presuppositions that it brings with it, then we simply cannot get from Fa to a falsehood, if we say $(\exists y)(Fy)$. If this is all it amounts to, then the principle of existential generalization cannot be invalid.

Note next, however, that to say what I have just said for existential generalization is not yet to say that we should expect

$$\Box(Fa \supset (\exists y)(Fy)).$$

For such a claim to be true, the conditional $(Fa \supset (\exists y)(Fy))$ has to hold with respect to every world. But the idea that it might hold with respect to every world is more problematic than the defense that I offered of existential generalization. For if $(Fa \supset (\exists y)\, Fy))$ is to hold with respect to every world, then one has to be prepared to assign this conditional the value true even with respect to worlds that lack the object *a*. Logicians have thought of ways to do this. But might one not prefer to avoid so evaluating such formulas for these worlds? And, if one does prefer to avoid this, then one ought to doubt the simple necessity of the conditionals that correspond to the inferences that are licensed by existential generalization. Let us affirm existential generalization then, but let us do so without rushing to affirm the necessitation of the corresponding conditional. All that existential generalization has to preserve is truth, not modal status.

13. Is there not something unsatisfactory, someone may say, about this refusal to assert the necessitation of existential generalization? Surely, if you hold that such generalization really is truth-preserving, then you have to say that it *can't* be that *Fa* without its being that $(\exists y)(Fy)$. Isn't some such modal claim inescapable? How then can there fail to be an entailment or strict implication $\Box(Fa \supset \exists y\,(Fy))$? My reply is that, when one says "it can't be that *Fa* without its being that $(\exists y)(Fy)$", that is indeed modal. But the modality in question is one that *presupposes* that there is such a thing as *a*. Perhaps then, allowing for that, one could say:

> It is necessary, on the presupposition that there is such a thing as *a*, that $(Fa \supset (\exists y)(Fy))$.

and this does indeed have the consequence that interests us, namely

> It is necessary, on the presupposition that there is such a thing as *a*, that $\sim(\exists x)(x = a) \supset (\exists y) \sim (\exists x)(x = y)$.

But as a whole, with this *a*-relative necessity, the necessity presuppositive of *a* (the presupposition itself conflicting as it does with the state of affairs adverted to in the antecedent of the conditional), this conditional says nothing inconsistent with the nonrelative or nonpresuppositional simple possibility of there being no such person as Julius Caesar.

Someone may be tempted to think that the presupposition of the necessity of a sentence can always be translated into a further condition to be transferred to the sentence as a new conditional antecedent. But they should test that claim carefully, and find a general argument for it, before they affirm it, first noting its formal or abstract affinity with the error of supposing that a conditional probability ought to be equivalent to the probability of a conditional. Until the parallel is disarmed, we should surely refrain from looking for any unconditionally or nonpresuppositionally modal counterpart for existential generalization.

14. The second difficulty, deriving from universal instantiation, invites a similar treatment. On the supposition of Caesar's existence, we cannot go from a truth to a falsehood in passing from 'everything is Φ' to 'Caesar is Φ'. But the 'cannot' is Caesar-relative or presuppositive of Caesar. No simple necessity, no necessity for all worlds, is shown thereby to attach to the conditional 'If everything is Φ, then Caesar is Φ'. Nor should it be automatically supposed that a Caesar-relative necessity can be transformed into a simple or nonrelative necessity by the affixing of a further condition to the conditional governed here by the Caesar-relative necessity.

15. How then is the simple or absolute contingency, which is not, I say, inconsistent with Caesar-relative necessity and which is claimed by option (III) for 'Julius Caesar exists' and Julius Caesar doesn't exist', to be marked positively?

Here we need to revert to our basic form (∃ *x*) (*x* = Julius Caesar) and single out the particular articulation favored by option (III). Then we need a notation that outruns the first-order forms, which fail to differentiate these articulations, in order to attach the contingency to precisely the thing that option III says it attaches to, namely the instantiation of the first-level concept that is the concept of Julius Caesar.

If our only way to identify that first-level concept is by an existentially committal reference to Caesar himself, our problem is akin to that which we have when we want to claim that it is contingent that the nastiest dog alive in 1993 is nasty. The solution in this case is to understand the definite description in Russell's way (as I should want to do in any case) and see the sentence in question as saying:

> (∃ *y*) (*y* is the nastiest dog alive & (∀ *w*) ((*w* is nastiest dog alive) ≡ (*w* = *y*)) & *y* is nasty).

Or, in binary notation with a "the" quantifier, we can write

> (The *y*) (*y* is the nastiest dog alive in 1993, *y* is nasty).

The thing that is contingent here is the content of the last clause, namely that *y* is nasty, not that whatever dog is nastiest is nasty.

Encouraged by the parallel, let us single out the first-level concept that is determined by '= [Julius Caesar] in our basic form '(∃ *x*) (*x* [= Julius Caesar])', and let us form a corresponding first-level concept designation

> λ*y* (*y* = Julius Caesar)

And then let us form the second-level concept that remains when '[= Julius Caesar]' is removed from the basic form with which we began, to obtain

$$\lambda\Phi\ (\exists\, x\ (\Phi x)).$$

This is the second-level concept of instantiation. Then, arriving at the stage we had reached when we first considered 'the nastiest dog alive is nasty', we can write

$$(\lambda\Phi\ (\exists\, x\ (\Phi x))),\quad \langle \lambda y\ (y = \text{Julius Caesar})\rangle.$$

But to be precise about what exactly is contingent – as we were with the nastiness of the nastiest dog – let us spell this out further as follows:

There is a first-level concept Ψ, such that Ψ is of Julius Caesar and, *contingently*, Ψ is instantiated

or

$(\exists\ \Psi)\ ((\forall\ w)\ (\Psi w \equiv (w = \text{Julius Caesar}))\ \&\ \textit{contingently}\ (\lambda\Phi\ (\exists\ x)\ (\Phi x)), \langle\Psi\rangle)$

or, in binary notation with a higher level 'the' quantifier,

$(\text{The}\ \Psi)\ ((\lambda y\ (y = \text{Julius Caesar}) = \Psi), (\textit{contingently}\ (\lambda\Phi\ (\exists\ x)(\Phi x)), \langle\Psi\rangle))$

16. Contingency being vindicated, we now arrive at the notorious problem of the true singular negative existential. But what exactly, for the friend of option (III), constitutes the problem?

Suppose that, referring to Leverrier's hypothesis about the perturbation of the perihelion of the planet Mercury, I say

Vulcan does not exist

or

There is a first-level concept that is of Vulcan and this concept is not instantiated.

And suppose I speak truly when I say this. Then, by existential generalization, which there is no evident way to escape, I am committed to the following as a truth:

There is an x such that there is a first-level concept of x and this concept is not instantiated.

But this is impossible. So, by contraposition, it seems it simply cannot be true that Vulcan does not exist.

Here the odd reflections into which we were led by the false singular negative existential may suggest the need for extra circumspection before we concur in the claim that there is a real difficulty. I declared that, if I said Vulcan didn't exist, then I was committed to something that was impossible, a fortiori false. But by what kind of necessity? The answer is that I was only committed by a Vulcan-relative necessity, a necessity presuppositive of Vulcan. This is a necessity that attaches to the use of the name Vulcan. But here I have no escape from that necessity. I need to use this name (or an equally problematic substitute for it) if I am to say what individual concept it is that I wish to declare to be uninstantiated.

That is the remaining problem – a problem it is tempting but ultimately hopeless, I think, to try to solve in Russell's way by resorting to some non–entity-invoking description that dispenses altogether with 'Vulcan' or 'Vulcanizes'. What seems certain is that we can get no help here from the expedients by which we defended the contingency of the false negative existential statement and the true positive existential statement. For here, in the true negative singular existential, we encounter the case of something we have to say if we want to speak truth, where this thing apparently commits us to nothing less than a falsehood.

17. In the face of this old difficulty, we shall not make any progress, I believe, unless we allow ourselves a change of scene. Or rather, we must take an interest in the historic background to the sad disappointment that is recorded by the words 'Vulcan does not exist'.

Until 1846 there was no theory of Uranus that permitted its movement to be represented satisfactorily. In 1821 Bouvard had contructed tables that adhered very closely to recent observations. Yet twenty years later a discrepancy of two minutes had already been observed. . . . Le Verrier demonstrated that [Uranus'] observed perturbations could not be explained as the effects of the actions of Jupiter and Saturn. . . . He began to search for signs of an unknown disturbing planet. Finally . . . on 31 August 1846, Le Verrier fixed the exact position of the unknown planet and gave its apparent diameter.

Le Verrier communicated the result of his investigations to several astronomers who had powerful instruments at their disposal. Among them was J. G. Galle, at the Berlin Observatory, who was notified on 23 September. Two days later, he wrote to Le Verrier: "The planet whose position you indicated *really exists*. The same day when I received your letter I found a star of the eighth magnitude that was not recorded on the excellent Carta Hora XXI (drawn by Dr. Bremiker). . . . The observation of the following day confirmed that it was the planet sought." The map to which Galle referred had just reached him. The object observed was fifty-two minutes from the position predicted by Le Verrier. First called Le Verrier's planet, it was eventually called Neptune. . . .

Le Verrier hoped that his work would lead to further discoveries analogous to that of Neptune. . . . He knew by 1843 that the observed motion of Mercury would not agree with theory. In 1859 he showed that Mercury moves as if an unknown agent produced an advance of its perihelion of about thirty-eight seconds per century. He then put forth the hypothesis of an intramercurial planet [which was later called 'Vulcan'] [or] of a group of such planets. The dimensions of a single perturbing body – on the order of those of Mercury – ruled out the possibility of this body remaining unobserved. The hypothesis was not confirmed."

Jacques R. Lévy, "Le Verrier"[13]

18. Let us begin with the success, which was Neptune. The first thought one will have in reading this part of the narrative is how suitable the claim "Neptune exists" is to be paraphrased à la Russell by a pure (not Neptune invoking) description, namely the description Le Verrier supplied to J. G. Galle, when he was asking him to search for the planet. At this time, Le Verrier was using no name of Neptune at all.

It is a defect of this proposal that it has no generality commensurate with that of the problem that we have with the negative existential statement that is true. But, before we return to that, we should also note that, strictly speaking, the object Galle found did *not* in fact match Le Verrier's description. Le Verrier had misdescribed Neptune's position – and he could surely have misdescribed it in various other ways without much detriment to the search Galle conducted. Will Russell's method enable us to dispense in the Neptune case and the Vulcan case, and in all the other cases for which these are our hostage, with the idea of a *particular object* that the Frenchman had conceived of or misconceived of and that in the happier case the German astronomer (nevertheless) found? If Russell's method does not enable us to do that, then the way in which we must understand which object the object is that gives the name 'Neptune' or 'Vulcan' its sense (and identifies the individual concept that is instantiated or not instantiated) has to be the same as the German astronomer's seems to have been in the Neptune case, namely this: We make ourselves party to Le Verrier's thoughts and his hypothesis. In becoming party to those thoughts, and speculating about how it's as if things are with Uranus and its neighbors in the solar system – and then in participating in Le Verrier's later speculations about how it's as if things are with Mercury and its neighbors – we are drawn into having thoughts of our own that make reference to the very same subject matter that Le Verrier had in mind.

Some will still insist on respecifying the concept of Neptune or Vulcan in Russell's fashion. Most of the difficulties we have been struggling with would disappear if we could generalize this solution to all cases where there is a temptation to predicate existence of an object, or if we could dispense with the Fregean cum Quinean move we were making at Section 4. But the reason why these questions are still alive is that, as I have admitted, the Russellian defense of the Frege–Russell view of existence is hard to generalize. And even in the present case, which is relatively favorable to Russell's method, is this not a misconception of how we understand Le Verrier's hypothesis? What the theory of reference seems to require is not the reduction of Le Verrier's speculative thoughts (or the reduction of the thoughts that we ourselves have as we follow Le Verrier's scientific argument) to other less committal thoughts, but a better way to understand the status and power that such speculations have in their unreduced prephilosophical condition.

Perhaps the most important clue to what is involved in this test is provided by Galle's words 'The planet whose position you indicated *really exists*' (Galle's emphasis). Under the option (III) which we are to redeploy, Galle's words are perfectly all right as they stand – but they are tantamount to this: 'You told me about, and you tried to specify by position etc., a certain planet I was to look for. You wrote as if there were a certain planet perturbing Uranus and whose position you were estimating. There really is ⟨of⟩ that planet; the concept really is instantiated'. And if he had been giving a verdict on the Vulcan case, he would have had to say: 'There really is not ⟨of⟩ that planet: the concept is really uninstantiated'.

Why the 'really'? Galle says 'really' because he finds that the set-up that Le Verrier has claimed that it is as if there is in the neighborhood of Uranus, including a certain planet that has a perturbing effect there, really can be found in that neighborhood. Le Verrier has full freedom to describe how it's as if it is, provided that he is arguing his way forward, not simply spinning words. He is free to sketch a possible set-up and free to identify things within that set up, employing under the auspices of the *as if* the ordinary terms we ordinarily employ to describe what we actually identify and observe. But all this is in the expectation (everybody's) that, in the end, Le Verrier will address the question, or persuade someone like Galle to address the question, whether the how it's as if it is and the how it is coincide. What 'really' marks is the transition – within one sentence – from one sort of thinking and talking, the speculative mode, to the other sort, a reality-invoking mode. What the true singular negative existential will force us to recognize is that we need both these modes, and that we need the means to negotiate transitions – even transitions within one sentence – from the one mode to the other.

19. The whole credit for any insight of this general kind must belong to Gareth Evans. See especially his p. 369

> The general idea is that someone who utters [a negative existential] should be likened to someone who makes a move within a pretence in order to express the fact that it is a pretence. He is not like someone who tries to prevent a theatre audience from being too carried away by jumping up on the stage and saying: 'Look, these men are only actors, and there is no scaffold or buildings here – there are only props'. Rather, he is like someone who jumps up on the stage and says: 'Look, Suzanne and the thief over there are only characters in a play, and this scaffold and these buildings

are just props'. The audience must be engaged or be prepared to engage, in the make-believe, in order to understand what he is saying.

But, in order to fend off misunderstanding, I mark now some of our differences of doctrine or of emphasis. These are differences that I think I do not attribute to the fact that Evans is an adherent of option (II) and I am an adherent of option (III), or to the fact that Evans ignored hypothesis and concentrated on fiction and I have concentrated here on hypothesis.

First, Evans chose as the background logic for his speculations a free logic. Now it may well appear that free logic will serve very well to underwrite the speculative liberty that we need to extend both to Le Verrier in the conduct of his speculations and thought experiments and to ourselves in outlining their content. But it does not suffice. For within the *as if*, existential generalization (a misnamed principle perhaps, but the move from Φa to $(\exists x)(\Phi x)$ and to $(\exists y)(y = a)$ is what I mean) is no less indispensable than singular reference. Indeed, to judge from Lévy's sort of narrative, one needs 'some planet', 'the planet' and 'that planet' more badly in this connection than one needs 'Neptune' or 'Vulcan'. Within such speculation, as outside it and everywhere else, what we need is ordinary logic with existential generalization.

In the second place, Evans sees one who goes along with a story or explanation (does not report it but goes along with it, rehearses it, develops it, etc.), and who does so in order to speak of the existence or nonexistence of the objects mentioned therein, as being in the state of *as if* understanding which objects these are. He says nothing stronger than that. But surely it's not merely *as if* we understand how Le Verrier says things are. Rather we understand, we actually do understand, how Le Verrier says it's as if things are, both in the Neptune case *and* in the Vulcan case. As Jacques Lévy says, Le Verrier showed in 1859 that Mercury moved as if an unknown agent produced an advance of its perihelion of about thirty-eight seconds per century. In this way Le Verrier argued that it was as if there were a planet that produced that effect. And then he went on to speculate about *that* planet. People would have known what he meant by 'that planet'. Indeed he asked colleagues who understood him to help him find it. Has not Evans failed then to get the *as if* into the right place? If we follow Le Verrier's argument in detail, and if we do not demur from his development of it, seeing nothing arbitrary or capricious in it, then we can understand as well as you like how it is as if things are, what it's as if Neptune is, and what it's as if Vulcan is. Our understanding can be just as good as understanding ever can be – of that kind of thing. If it weren't good enough, no 'game to reality shift' could ever improve matters.

20. Back now to 'Vulcan does not exist' and the problem that it makes for us. (This is a problem we could equally well have had with 'Neptune doesn't exist', but which we don't in fact have with that sentence.) According to option (III), it is simply contingent that the thing we mean by saying 'Vulcan does not exist' is true. But it seems that what we normally mean by the sentence is not *best* said by the words 'Vulcan does not exist'. It cannot be. For the use of the name 'Vulcan' depends on a presupposition, namely the availability of the bearer and its determinate identifiability, that the whole message apparently contradicts. (That point has been made a million times before. It is none the worse for the repetition.) What we would do better to say, in any full or official version, is 'Vulcan doesn't *really* exist',[14] where the 'really' precisely signals the necessary transition from the *it's as if it is* to the *is,* the transition from reference to what it's as if there is, including a planet affecting the perihelion of Uranus or including a planet affecting the perihelion of Mercury, to the matter of the actual instantiation or noninstantiation of the concept of that thing. 'As regards that planet, the planet which (it's as if) affects the perihelion of Mercury, and as regards *its* concept *c,* well, really nothing at all answers to *c, c* is not really instantiated' where the scope of the 'really' is similar to that which we assigned earlier to 'contingently'. (See §15, ad fin., above. See also note 14.)

21. On this rendering, spelled out in the official language of option (III), we find ourselves with one foot in Le Verrier's speculation about the perihelion of Mercury and the disturbing influence of Vulcan, this speculation determining the concept of Vulcan, and with the other foot in the real world, saying that in reality nothing answers to that concept. (This seems to be the only pattern that can be generalized to cover all cases where the true negative existential statement causes trouble.) But is that not an uncomfortable position? If we want to see the sentence 'Vulcan doesn't exist' as a straightforward common or garden truth, then we shall itch to pick up the first foot and put it down with the other foot on the terra firma of *real, really, reality.* We think we want to be able to say not merely 'Vulcan doesn't exist', or as regards Vulcan there isn't really ⟨of⟩ it'' but 'Really Vulcan doesn't exist' or 'Really, as regards Vulcan and the concept of it, there doesn't exist ⟨of⟩ that concept' with maximum scope for the 'really'. We think we want to be able to say these last things, but without, of course, committing ourselves to their apparent consequence, namely that really *there is* something to the concept of which nothing answers.

To secure the possibility of saying these things, with a wide scope 'really', but without having to say that there *really* is this object, we might try sealing off the reference to the object inside an 'as-if' clause, thus

There is a concept *c* such that its as if *c* is of a certain planet such that . . . [where . . . is the whole of Le Verrier's speculation], but nothing answers to this concept.

Such a claim so formulated, with the 'as if' insulating all reference from the larger context, might seem to support a 'really' prefixed with wide scope. But what is the concept *c* the concept of? What concept really is the concept *c?* We seemed to be almost home. Yet we aren't there. By this route, how could one get any nearer?

22. Another way we may try to advance is by back sliding into some principled adjustment of the doctrine we have accepted of singular terms, and/or some relaxation of the commitment to existential generalization. Someone might so tinker with these things that we were able (he said) to assign plain truth-conditions to 'Odysseus was found by Nausicaa' (not true but 'true in fiction'?) or 'Vulcan disturbs the perihelion of Mercury' (false) and so forth.

I have already said my piece against all this and will add only the following. It obscures the fundamental fact that the 'as if' of fiction and hypothesis is always *modeled* on the 'is'. In the story, or in the hypothesis, we try to get as close as we can to a point where one can think of each object as being arbitrarily specifically or determinately identifiable within the space of the narrative or hypothesis. It is a topsy-turvy procedure to rebuild general logic and the logic of fact in order that we should allow there for fiction, if fiction itself aspires to the condition of as if *fact.*

23. The other thing we can try to do is to learn to live with existential generalization, acquiesce in the necessity that the true singular negative existential makes for us to leave one foot where it was (namely in the territory of storytelling or speculation, where it needs to be for us to determine which concept Vulcan's concept is), while leaving the other foot where it wants to be, in the territory of the real (the place where we are when we assert that really nothing answers to the aforesaid concept). In the case where the singular negative existential is true and there really is no such object (the Vulcan case), the concept in question is determined by the *as if* and it will never ever be able to be determined in any other way. But that does determine it.

The account that emerges of saying 'Vulcan doesn't really exist' (or 'Odysseus does not really exist') is then as follows. To say this, we have to rehearse the hypothesis yet again (or tell the story yet again) but, this time, in the course of the rehearsing (or the telling) and the further developing or amplifying, we must make a comment. We must say that the hypothesis fails, or say the story is only a story. But we do this *in the course of the rehearsing (or telling) it.* When we are in that rehearsing mode and are switching to the other commenting mode, we shall not bother to resist if someone says to us: 'So you are prepared to say that there is something that doesn't really exist'.

So long as 'really' has the small scope we have explained, we do not need to resist.

If all this is right, then we do not, when we issue a true negative existential, move the hypothesis or the story into the actual world and register its failure there. That would be lifting up the foot that rests in the world of speculation in order to put it down with the second foot on the *terra firma* of reality. We can't do that.[15] Just as there is no better or cleaner way of saying what is said by 'Look, Suzanne and the thief over there are only characters in a play, and, this scaffold and these buildings are only props.'

24. If we do in this way leave proper names where they were, leave existential generalization where it was, and leave logic where it was, then it will come to worry us whether stories and unsuccessful hypotheses have a completely determinate sense. But to say that they don't is only to say (it's to say no more than) that our comprehension of what hypothesis and story involve falls short (and must from the nature of the case fall short) of the (indefinitely and open-endedly improvable) comprehension one can have of something that is unproblematic, true, and unproblematically true. This falling short of that ideal does not mean that we *don't* comprehend hypotheses or stories – or that, at best, it's only *as if* we comprehend them.[16]

On these terms, how can the negative existential statement count as straightforwardly true? To that I answer that the sentence that expresses it is not a straightforward kind of sentence. The grounds on which we are likely to utter it will be that the object-containing hypothesis or story is a hypothesis or a story and no more: that one can't find anything in the real world to match up with the object in the story. If we are dismayed by the point that, strictly speaking, the materials just furnished do not recapitulate (or confirm the availability of) ordinary compositionally generated truth/falsity conditions, then it is likely we may come to see a new point in Frege's and Russell's however mismanaged or discredited attempts to deny that sentences like 'Pegasus exists', 'Vienna exists', and so on, can make sense. Perhaps such sentences as these should be *replaced*, someone may reflect – for example, by our *showing* the actual things, all of them, that are there to be found in the world, none of them answering to a Vulcan- or Odysseus-specification. Or perhaps one ought to look with a new interest at the question whether reality can be adequately characterized without any recourse at all to such forms as the singular negative existential. But I am not sure that one needs to assist actively in such inquiries as these. What we actually do, and apparently have to do, in order to arrive at the utterance of a true singular negative existential statement is so much more interesting than any of the alternative things that might be done by a being infinitely better informed than we are and infinitely more literal minded.[17]

NOTES

1. For Frege's hierarchy of objects, concepts etc., see Michael Dummett, *Frege: Philosophy of Mathematics* (London: Duckworth 1991), pp. 89–93. For a defense and a minor but liberating amendment of the doctrine in the face of paradox of the concept *horse*, see my "On the Sense and Reference of Predicates, etc." *Philosophical Quarterly* 34 (1984). For the further elaboration of the Fregean doctrine of concept-words in application to the understanding of the mechanisms of natural language, see my "Putnam's Doctrine of Natural Kind Words and Frege's Doctrines of Sense, Reference and Extension," in A. W. Moore (ed), *Meaning and Reference* (Oxford: Oxford University Press, 1993).
2. Compare Andrew Rein, "Exist(s)", *The Philosophical Quarterly* 38 (1988): 471–80. On the question of whether this difference amounts (as I suppose) to an ambiguity or to something in which it would be easier for a general account of the sense of 'exist' to acquiesce, see S. G. Williams, "Existence," forthcoming in *The Blackwell Companion to Metaphysics*, ed. Ernest Sosa and Kim. On the question of ambiguity, and what hangs on it, see also Timothy Williamson, "Equivocation and Existence," *Proceedings of the Aristotelian Society* 88 (1987–8). The tame tigers example is of course G. E. Moore's. See "Is Existence a Predicate?" *Proceedings of the Aristotelian Society*, supp. vol. 15 (1933): 175–88.
3. See Gareth Evans, *The Varieties of Reference* (Oxford: Oxford University Press 1982), chapter 10.
4. See p. 50 in P. T. Geach and Black, eds., *Philosophical Writings of Gottlob Frege* (Oxford: Blackwell 1951).
5. Similar considerations will appear when we seek to proscribe the inference from 'tame tigers are few and far between' and 'Leonard is a tame tiger' to 'Leonard is few and far between.' Cp. Bertrand Russell, "The Philosophy of Logical Atomism," in *Logic and Knowledge*, ed. Marsh (London: George Allen & Unwin, 1956), p. 233.
6. This is a provisional account, and other possibilities will emerge. See §24 of this paper.
7. How then do we say that there *is* or *exists* such a concept as the concept of tame tiger? Presumably by saying that there is ⟨of⟩ the concept of the concept of tame tiger – or that the concept of the concept of tame tiger is instantiated.
8. For "Caesarizes", etc., see W. V. O. Quine, "On What There Is," in *From a Logical Point of View* (Cambridge, Mass: Harvard University Press, 1953). For some of the difficulties of such strategies, see Donald Davidson, "Theories of Meaning and Learnable Languages," in Y. Bar-Hillel, ed., *Logic, Methodology and the Philosophy of Science* (Amsterdam: North Holland, 1965).
9. For this conception of the meaning of singular terms as it has evolved over the last three decades, culminating in McDowell's proposal exploited in the text, see G. E. M. Anscombe, *Introduction to Wittgenstein's Tractatus* (London: Hutchinson, 1959), pp. 41–4; Michael Dummett, "Truth," *Proceedings of the Aristotelian* Society 59 (1958–9), p. 153; Ruth Barcan Marcus, "Modal Logics I," *Synthèse* 27 (1962): 303–22; David Wiggins, "Identity Statements," in R. J. Butler, ed., *Analytical Philosophy* (Oxford: Blackwell, 1965), pp. 43–6; John McDowell, "On the Sense and Reference of a Proper Name," *Mind* 86 (1977); David Wiggins, "Frege's Problem of the Morning Star, etc.," in M. Schirn (ed.), *Studies in Frege*, vol. II (Stuttgart-Bad Constatt: Fromman-Holzbog, 1976); Gareth Evans, *Varieties of Reference*, chap. 2.
10. Cp. Rein, "Exist(s)."

11. See P. T. Geach, *Reference and Generality* (Ithaca, N.Y.: Cornell University Press, 1962).
12. See Saul A. Kripke, 'Semantical Considerations on Modal Logic,'' *Acta Philosophica Fennica* 16 (1963): 83–94.
13. ''Le Verrier,'' in vol. 7 of *The Dictionary of Scientific Biography* ed. Charles Gillispie (New York: C. Scribner's, 1981). For the thought that the problems of speculative reference deserve attention equal to that accorded to fictional names, see R. M. Sainsbury's invaluable review of Evans, *Varieties of Reference*, pp. 138–41, *Mind* 93 (1984).
14. Cp. Evans, *Varieties of Reference*, p. 352. Given Evans's and my different attitudes to free logic, it will not be open to me to allow this 'really' the wide scope that Evans will claim for it. See §22 of this paper. Nor am I seeking here to recapitulate the details of Evans's view, which is shaped in part by his choice of option (II).
15. If we feel we must have both feet in the same place, we can try the opposite tactic, move the other foot from terra firma and put it down with the first foot in the realm of as if. Perhaps that is what we have to do if we want to say that Prince Albert did *not* marry the Lady of the Lake. In this case we must move Prince Albert into the story of the Lady of the Lake and find that he is no bridegroom there to that Lady. We cannot move the Lady of the Lake into the real world and try to frame there the true proposition that she was not a bride to Prince Albert. Among real things, we cannot make sense of the question of which thing she is.

    This suggestion may seem as fantastic as Mahomet's claim that the mountain was going to move to Mahomet, not Mahomet to the mountain. (Should I try to advocate a similar treatment of 'the Greeks worshipped Zeus'?) But consider the apparent ease with which you comprehend the striking sentence 'some ([*sŭm*] not [*səm*]) dragons exist – or better 'some [*sŭm*] dragons really exist'. (Cp. Richard L. Cartwright, *Philosophical Papers* (Cambridge, Mass.: MIT, 1987), ''Negative Existentials,'' pp. 21–31.) To understand this sentence, reflect that here we are not simply telling a story, we are epitomizing *many* dragon stories – The Loch Ness story, the story of the Ring Cycle, the story of St George, etc. – and then we are saying that, among the items mentioned therein, there are *some* (e.g., Fafner) such that, there really is ⟨of⟩ them. Their individual concepts are really instantiated. If so, there is a real object in Wagner's story.

    With this sentence we are not saying what we should be saying if we declared ''some [*səm*] dragons exist'' where we might as well have said 'dragons exist''. That's not what '[*sŭm*] dragons really exist'' means here.
16. This is to claim that we comprehend them *well enough*. To see the indispensability, for any satisfactory theory of these matters, of degrees of comprehension, consider the paradox that is raised at Plato *Meno* 80^{d2}. (If degrees of comprehension are irreducible, are degrees of intelligibility irreducible too?)
17. At one time and another, I have greatly benefited from talking about the question of existence and some other questions touched upon in this essay with Christopher Peacocke, Andrew Rein, Timothy Williamson, Peter Milne, Mark Sainsbury, and John Campbell.

Postscript

Comments and questions directed to me since the submission of this essay suggest that the chief contentions of §§21–25 stand in need of a brief summary. I offer the following, in full awareness that such a resumé can only stir up new questions.

1. In the case of 'Vulcan', one knows which thing it is as if it refers to. ('It's as if p' entails neither that p nor that not p.)
2. For in the case 'Vulcan', one knows that it's as if 'Vulcan' refers to Vulcan.
3. That one knows that it is as if 'Vulcan' refers to Vulcan – this is something that the theorist explains in seeing our thoughts as downwind from Le Verrier's and explaining Le Verrier's thoughts by entering into his hypothesis and making a hypothesis-relative identification of Vulcan, by reference to the question of Mercury's perihelion etc. This identification is achieved in what Evans would call the conniving mode – a mode the theorist needs as badly as everyone else.
4. The conniving mode of explaining adverted to in (3) above is indispensable. This mode of explaining admits such comments as 'there isn't *really* ⟨of⟩ / there *really* is ⟨of⟩ the concept *identical-with-Vulcan*'.
5. So long as human beings make up hypotheses or stories, speakers and the theorists of language who shadow these speakers will need the conniving mode of explanation of reference. So long as we remain in touch with reality itself, however, the conniving mode will itself need the option to make comments such as those mentioned in (4) above. Comments of this sort are offered *within* the conniving mode, directed *upon* what is being attempted there, but answerable to reality.

14 January 1994

II

Morality

8

Moral Dilemmas Revisited

PHILIPPA FOOT

I have called this paper "Moral Dilemmas Revisited" to mark continuity with a paper I published ten years ago under the title "Moral Realism and Moral Dilemma."[1] The point of that piece was to challenge an argument against moral realism put out by Bernard Williams in the sixties, in two articles: "Ethical Consistency" and "Consistency and Realism."[2] Moral realism is not the topic of the present paper. But I want to look again at moral dilemmas. Ruth Marcus was one of the first to arouse my interest in the subject, and it seems fitting to offer this paper for her *Festschrift.* She would, I think, be the last person to want the affection and respect of her friends to take the form of agreement rather than challenge; so I hope I shall be offering her something she will like.

Marcus's views on moral dilemmas appeared in an article called "Moral Dilemmas and Consistency," where, very controversially, she insisted that it is possible to be in circumstances in which one will be guilty whatever one does: "damned if one does a certain thing and damned if one doesn't" as she says.[3] This is the thesis that I want to oppose. I do not agree with her that the feelings of guilt that people often have when they have had to make a difficult moral choice imply that they *are* guilty; or indeed that we can even understand the possibility of inescapable guilt, unless we believe, for instance, in a cruel God who damns us for disobeying either of two laws on occasions where both cannot be obeyed.

With these doubts I shall also once more be querying a thesis put out by Bernard Williams, who, like Marcus, thinks that feelings of guilt should be taken very seriously, and who has erected upon such feelings a doctrine of the possibility of an agent faced with inevitable wrong action. How, asks Williams, can we rule this out? What facts about the world or about an agent could do so?

> . . . There is a substantial and interesting question: "What would have to be true of the world and of an agent that it should be impossible for him to be in a situation

Special thanks to John Campbell and Gavin Lawrence for help with this paper.

where whatever he did was wrong?'' I doubt in fact that there is anything that could produce such a guarantee short of the existence of a rather interventionist God, or else the total reduction of moral life to rules of efficient behaviour–two extremes which precisely leave out the actual location of moral experience. But it is at any rate a real question, and it would not be a real question if the correct thing to say were that nothing has to be true of the agent or of the world for this to be so, because it is guaranteed by the logic of moral expressions.[4]

In the end I shall focus on Williams's view, which seems more basic than that of Marcus, in that her thesis requires the truth of his. For she wants to say not just that feelings of guilt are natural and even creditable after a hard choice, but also that the subject who has them may actually *be* guilty whichever horn of a dilemma he chooses to grasp. And for this it is not enough that he can't get away from the idea (and so from the feeling) that he has done something wrong; it must be true that he has.

Marcus's position is quite extended: Since she thinks that the agent's choice must leave him just a little guilty even in the case of a moral dilemma in which ''priorities are clear and overriding.'' She says that in such cases the guilt is ''appropriately small.''[5] But why does she think that there is any guilt at all, when, for example, someone justifiably breaks a promise in order to fulfill a more pressing obligation? What she is really interested in is, it seems, not guilt but rather ''remainder'': She wants to deny that an obligation that is overridden simply disappears. So we should look at the doctrine of ''remainders'' with respect to moral dilemmas, found not only in Marcus but also in Williams and others, and see what, if anything, it has to do with guilt.

A moral dilemma can be characterized, first of all, in terms of evidence. Seen in this way, a moral dilemma (which as understood here need not be such as to put anyone in his senses into any uncertainty as to what to do) will have to be a special case of the dilemma that exists wherever there is evidence for and evidence against a certain conclusion. What is special is that the conclusion is about what an agent ought to do. The relevant ''ought'' signifies what Williams has felicitously called ''the practical *ought*,'' which he has also called ''the *ought* of advice.'' Moreover it speaks of what should be done not from some special point of view, but rather all things considered; and gives a verdict, where there is a verdict to be given, about which of two conflicting principles overrides the other. In the present context it may be thought of as ''the moral *ought*,'' so long as moral principles can be assumed both valid and overriding. One should note, however, that in some other discussions this would be in question, as when Thrasymachus, giving advice, would say that anyone strong enough to get away with injustice *should* try to do that.

Let us represent this practical, all things considered, *ought* with the formula OXA for 'X ought to do A'. And now let us consider the evidence or grounds for such a judgment. These grounds are expressed in a second class of propositions, stating, *inter alia*, the existence of rules, promises, engagements, and considerations about good and harm to oneself and others. There is also a third class of propositions that assert an evidential relation between such things and the "practical *ought*" (i.e., which assert their status as grounds). So, still, for the moment, speaking of evidence, we say, for example, that where X has promised to do A (call this consideration C) there is reason to believe that OXA: or 'Prima facie, on C, OXA'.

What is confusing here is that there is an ambiguity in 'ought' which allows a proposition about what an agent ought to do to be either of the type to which we have so far assigned it, in which it is Williams's "practical *ought*"; but also, along with assertions about the existence of obligations, promises, rules, benefits, and so forth, to appear in the propositions that serve as grounds. The term 'obligation' is even more ambiguous; it being quite natural to use this word not only for the all things considered verdict in moral contexts but also to mark, for example, contracts and claims engendered by special relationships. John Searle has suggested that it will be useful in this discussion to restrict 'obligation' to this use, while always understanding 'ought' to signify the "practical *ought*."[6] Alternatively we could flag the 'ought' that speaks of a *ground* for the practical *ought* in representing it, as well as Searle's *obligation* as O′: when we shall have OXA, O′XA, and Pf. OXA/O′XA.

What is the connection of all this with reasons for action? We notice, firstly, that there is an evidential relation parallel to the ones we have been considering, but this time between there being a reason for X to do A and there being best or conclusive reason for the same thing. So that we have RXA (best reason), R′XA, and Pf. RXA on R′XA. Moreover, since 'O' represents the practical *ought* of advice, OXA implies that there is, all things considered, best reason for X to do A. It is evident, however, that in making this explicit connection between OXA propositions and reasons for action, and in characterizing O′XA propositions as their ground we have tacitly restricted the O′XA range to propositions that themselves give reasons for acting, or at least give evidence for the existence of such reasons.[7] It is open to anyone to question whether such things as promises, family connections, or considerations about the good of oneself or others, really do give reasons for action and therefore evidence for propositions of the OAX form; but this is not my problem here. I shall proceed on the assumption that it has been solved, whether by showing that rationality is not a morally neutral notion, or by succeeding in joining a neutral concept of rationality to a standard, or fairly standard, moral code.

I shall therefore suppose, for the purpose of this discussion, that there are true propositions of this kind, represented in the formula OAX, and naturally read 'X ought to do A' (or 'should do A') where this implies that there are conclusive or best reasons for X to do A. I shall represent an alternative to A by the letter 'B' with subscript '~A', meaning to cover both the case in which B is an action specified by a description other than ~A and others in which this is its only specification. Thus I can speak alternatively of the dilemma about whether to do or not do A, or about whether to do A or B. In the latter case A and B may be consistent in the sense that at least in some possible world they could both be performed by a single agent on a single occasion, although they are such that in our world they cannot, in the circumstances, both be performed.

On this supposition we have, then, a conflict of reasons for action: in the sense that there are actions A and $B_{\sim A}$ and one reason at least for the doing of A and another for the doing of $B_{\sim A}$. And with the explicit introduction of reasons for action, as opposed to evidence for a proposition about what ought to be done, we have reached the place at which many philosophers have argued, rightly I think, that there is a special 'remainder' when a conflict of reasons has been settled either in the way it is settled when one reason or set of reasons has been found stronger than the other, or in the way it is settled when the dilemma has seemed insoluble and the agent has simply decided how to act: the suggestion being that this kind of remainder is not found in cases of evidence for a merely theoretical conclusion.

Considering 'remainder' I shall think first, for the sake of simplicity, about resolvable moral dilemmas: that is moral dilemmas where the evidence for OXA is stronger than the evidence for $OXB_{\sim A}$; where an obligation to do A would override the obligation to $B_{\sim A}$, and there is stronger reason for doing the first action than the second. The familiar old example is that of someone who has promised to be at a certain place to meet a friend but who has instead to drive an accident victim to hospital. Obviously he or she ought, all things considered, to do the driving rather than the meeting, and the question is what 'remains' from the obligations, reasons, and so on, on the other side.

First of all, let us consider 'linguistic remainder'; that is, what can we say in such a case if employing the same terms as those appearing in the O'XA or 'grounding' proposition. With some considerations this is simple. A promise is not annulled when its practical force is overridden: and when one has 'double-booked' engagements, the one that it is right to break is not any the less an engagement for that. These bits of language clearly work this way, but it is not so obvious in the case of obligations, and certainly not in that of reasons for action. Perhaps it is right for obligation, for it seems to make sense to say that one has an obligation that one cannot fulfill, and on that principle perhaps one can say that one has an obligation to do $B_{\sim A}$ even after the doing of A has made this impossible, so long as the time for doing it has not passed. But this does not seem to be true of reasons for action. For while it is right on occasion of conflicting reasons to say that "there is a reason" for doing each action (though not, of course for doing both) whatever the relative strength of the reasons may be, it does not follow that one still has a reason to do A when $B_{\sim A}$ *has been done*. This is partly because the other action is no longer possible and it is not clear that there can be a reason for an agent to do what he cannot do, and partly because the doing of one action may have destroyed the desirability of doing the other. Williams has pointed this out, with an example having to do with the desirability of courting, or marrying, either of two women but not both. And this is, of course, an extra reason, beyond the principle that *ought implies can* for dropping what he has called the principle of agglomeration in this area.[8] $O'XA$ implies that X can do A, and $O'XB_{\sim A}$ implies that X can do $B_{\sim A}$; but as A and $B_{\sim A}$ are incompatible, we cannot factor out to get O' (A and $B_{\sim A}$). And similarly there may be reason for doing each but no reason for doing both. So when one of the actions has been done there may no longer be any reason for doing the other; not, or not only, because one can have reason for doing only what one can do (if that is true) but also because the doing of either action (the better or the worse) may always have destroyed the reason for doing the other.

For some of these 'remainders' one may, therefore, have to take time into consideration when asking what effect being overridden in the practical or best reason stakes can have *on the original grounding propositions*. In some cases they can still be asserted, with a change of tense only at some time explicitly or implicitly referred to in the description of the action, but in other cases the doing of the alternative action may mandate a change of tense.

None of this is, I think, very interesting compared with other reasons that have been given for the presence of 'remainder' even in cases of resolvable moral dilemma. For instance, Susan Hurley believes that the possibility of *akrasia* shows that reasons for acting which are overridden do not *ipso facto* disappear; on the ground that in *akrasia* someone *acts on* a reason that they know to be other than the best.[9] This seems to me interesting but not quite right, on account of a distinction that I think must be made between what one might call (a) explanatory and (b) validating reasons, and the fact that an explanatory reason is enough (if one is indeed needed) for the possibility of *akrasia*. Consider, for instance, someone who does something for the sadistic pleasure that it gives him. Does he have a reason for what he does? In one sense he does, because there is a particular kind of explanation of the action: in terms of what Anscombe has called a "desirability characterization."[10] If he says this was the reason why he did it this gives his action a particular kind of intelligibility, and we could speak of "his reason" here. But we ought also to say, surely, that the fact that it will give the agent sadistic pleasure is no reason–not any reason–for doing it, as he himself may admit. Indeed, as Gavin Lawrence has pointed out to me, the fact that he will get pleasure from doing something may rather seem a reason against doing something than even the slightest, most overridable, reason on the other side; and he himself may even believe that.

With the existence of a reason for or against doing an action we are in the area of validating, or as I should like to say (in non-Davidsonian terminology) "rationalizing reasons"; that is, reasons having to do with the rationality of the action, not merely with its explanation.[11]

I do not know whether this distinction will be readily accepted. If not, this may be because the question of rationality can come in in a derivative way once an agent has determined on the pursuit of any end, however bad that end may be. There can be calculation and therefore practical reasoning about how to achieve any end, child abuse for instance, and a lack of rationality *relative to the pursuit of this end.*[12] But this does not mean that the agent even thinks there is reason to pursue the end. If he calculates well and follows where the calculation leads him, he does indeed avoid one failure in rationality; but there could nevertheless be *no* reason to pursue the end he is pursuing, and strong reason against it. His desires have no doubt played a part in determining what he does, and he may have held a 'desirability characterization' in his mind when he was taking the decision about what he would do. He lets it affect him in a way that is familiar to us all, particularly when we want to do something we know we ought not to do. And the practical *ought* may also determine our action. So three different stories are told here. There is the story about influences, in which the all things considered *ought* is (regrettably) only one competitor among others. Then there is the story about the weighing of reasons, in which the practical *ought* might seem to, but does not, have a speaking role: since it is not a pro among the cons. And finally there is the answer to the question 'What ought I to do?' in which no character except the practical *ought* is on stage.

The distinction between rationalizing and explanatory reasons is important, but I want it at the moment only in order to show why I think that Hurley is not quite right in saying that the possibility of *akrasia* shows that reasons for acting remain even when they are overridden by stronger reasons on the other side. For it *could* be (though it no doubt isn't) the case that all akratic actions were of the sadistic pleasure or child abuser's type; where "his reason" means only the 'desirability characterization' that makes what he did intelligible in a particular way. I agree with Hurley and others who have said that there may be proper (rationalizing) reasons for doing both of the incompatible actions A and B, and of course in many akratic actions there are these. This seems to me to be enshrined in the very ordinary notion of the pros and cons of a given course of action. For it isn't right to say "the supposed pros and cons'' with the idea that either the pros or the cons are going to disappear if the conflict can be resolved.

It makes good sense, therefore, to say that there are, for a given agent in given circumstances, reasons to do A and reasons to do $B_{\sim A}$, even when one set of reasons is stronger than the other. But it is not easy to see what more should be said about 'remainder' for moral dilemma in general. Those who stress this remainder are apt to talk about the moral requirement for apology and restitution to someone whose interests have been set aside. But this cannot always be in question, since a moral dilemma may involve only the agent's own life. Moreover, even where others are involved, as in the case of promises, it does not seem obvious that we owe anything, except an explanation, to another when we have *justifiably* broken a promise to him. Nor need our broken promise even have caused him any inconvenience, as we can see by changing the old example to make the circumstances of the two sides symmetrical. Now there are two accidents, and we meet, each with our accident victim, at the door of Casualty, not even having to explain why we are not where we promised to be. In this case noone suffers any inconvenience from the broken promises, and we can suppose that for some reason things turn out better for each of us than if the other's promise had been kept. (Perhaps a bomb went off in the building in which we were to meet.)

All of this shows that the supposed demand for apology and restitution will not serve to fill out the idea of 'remainder' for moral dilemma as such. Especially in the case of overridden promises, there may be nothing to it except that we would have kept our promise if we could have done, and that it would have been incumbent on us to look around for a way of removing the dilemma if we could; perhaps asking another driver at the scene of an accident if he or she were freer than ourselves.

Nevertheless, Williams was surely on to something important when he wrote of ''moral cost'' and ''moral loss'' in connection with the choice of one horn rather than the other in a moral dilemma.[13] This does seem appropriate at least for many cases, where it is not possible to winnow out the *loss of good* as in the previous example. Isaiah Berlin has persuaded us to take very seriously the idea of an inevitable *loss of value* – the necessity of sacrificing one *value* to another – both in political decisions and in an individual's choice of lives.[14] Berlin writes very naturally in this context about what we might call *human goods* such as life, health, liberty, and knowledge. And one notices that this natural way of talking can form a basis for assertions to the effect that it is good (or bad) that such and such happened, or is the case. Saul Kripke once gave me, when I was querying the objective use of such locutions, the example of the preservation and discovery of the Dead Sea Scrolls. This preservation and discovery surely was an excellent and important thing, and in saying so we do not simply reflect a personal wish or concern, any more than we would in speaking of the preservation and discovery of some great piece of music or of a great scientific discovery. A whole area of impersonal utterance belongs here: including all that we spontaneously say about 'great' or 'marvelous' things, about what is 'deep', and about 'the things that matter'.[15] The idea of *value* will come better into focus if we explore such utterances rather than the bare 'good' and 'bad', which, on their own, suggest Moorean intuitions of goodness and badness in happenings or states of affairs.

There is, therefore, no doubt but that Williams is right in thinking that there is something we might call moral loss or moral cost, at least in the cases of moral dilemma that Berlin has stressed. For we may have to sacrifice one thing that we value for the sake of another, such as family for achievement, or, as is only too frequent in times of war, great architectural monuments for lives. And indeed most moral dilemmas concern what are intelligibly called human goods, such as health, life, liberty, and knowledge.

It is not so clear, however, that this will go any way to vindicating Williams's idea of inescapable wrong action and therefore (possibly) Marcus's idea of inescapable guilt, because it does not seem that the 'badness' comes in the right place. To see that, one might consider as an analogy dilemma in musical performance. What does it mean to say that a musician played in the wrong way? There has got to be something bad about his performance; but obviously it is not enough that it was bad that he played as he did, since we might say that when a passionate pianist brings down the ceiling with his rendering of Rachmaninov. Nor, more relevantly, is it enough that there is some value, and moreover a musical value, sacrificed in his playing, as for example, pace for expressiveness, since this might be the very best way to play the piece. And it makes no difference if we suppose an insoluble musical dilemma, where there is equal or incommensurable loss in each of two incompatible renderings: and one has, as Berlin has put it, to "plump" for one or the other.

Let us turn, therefore, to another way in which Williams has tried to fill out the concept of *remainder;* that is with the idea of 'moral regret.'[16] At first sight it may seem that this is simply the psychological counterpart of 'moral loss,' and so something already recognized. But investigation shows that this is not right. It is true that in a few rather formal contexts regret may be for anything seen as bad, as when the bank manager regrets to inform us that we are overdrawn. But these are somewhat peripheral uses. Normally, regret is spoken of only when a choice or other action is in question. 'Do you regret your divorce?' implies that one chose the divorce or at least consented to it and now perhaps wishes that one hadn't. If one is burgled, one is sad about having been burgled; but it is the burglar, if anyone, who regrets the burglary, whereas what one oneself regrets is carelessness in locking up.

Moreover, while sadness can be about an action necessary to implement a choice that one thinks was all things considered for the best, this is not true of regret. So acting for the best in a moral dilemma, while it can entail sorrow, and in serious cases even horror, does not, if we have no doubts about the rightness of the action, make a place for regret. It is when there is no clear answer (and perhaps no answer at all) to the question of what we should do that we are likely to waver, so that some days we regret our action and some days we do not. Possibly it is on this account that those who stress 'remainder' are apt to pay otherwise puzzlingly special attention to irresolvable dilemma or otherwise difficult choices. This special attention is found even in Marcus, although her thesis is supposed to hold for all moral dilemmas.

For Williams it is essential to give special attention to what he and others have called "tragic choices," where a high degree of seriousness is combined with uncertainty, because it is here that he locates the possibility of inescapable wrongness; seeing this as something that the agent is faced with only in very exceptional cases, as when Agamemnon must sacrifice his daughter or lose his army. This "wrongness" is supposed to be something more than the "moral disagreeableness" of dirty hands – the dirty hands that both Williams and Stuart Hampshire, following Machiavelli, have described as being hardly avoidable in politics or in war.[17] There is no difficulty in understanding this "dirty hands" thesis, since even "dirty tricks" are sometimes justified, as they surely were when the Resistance fought the Nazis; and I suppose there can be no dirty tricks without dirty hands. But then those with dirty hands did not act wrongly, even though Williams seems to think that even here an innocent victim could complain that he had been *wronged*. Has he really been wronged if harmed, even if it was a *justified* action that harmed him? I should have thought not, because if the action *was* justified then if he does not recognize that he is making a mistake. It will no doubt often be difficult to get someone who has been harmed or 'used' to recognize the justifiability in a debatable case. But if he agrees with us that our action was justified, and even morally mandated, he must agree that we had a right to do it, and therefore, one would have thought, that he was not wronged.[18]

It is, in any case, in the very special situations – of terrible and insoluble dilemmas – that we are supposed to find the possibility of inescapable wrongness that Williams insisted on in the passage I quoted at the beginning of this paper. But even here it is hard to understand what kind of *wrongness* Williams has in mind. It is clear that 'wrong' could not there mean what it usually does. For suppose an insoluble moral dilemma of the kind that, according to Williams, leaves an agent with the choice between two wrong actions A and $B_{\sim A}$; and then a change by which one of the alternatives – say A – has been made even more ghastly, so that $B_{\sim A}$ is clearly the right thing to do. Then either $B_{\sim A}$ is still wrong, with a wrongness that is not incompatible with rightness; or else its 'wrongness' did, in some strange way, come from the fact that the old dilemma was irresolvable: in spite of the fact that genuine irresolvability seems to give us moral leeway to do either of two actions, rather than to confer a special kind of 'wrongness' on each.

What can it be, then, that makes philosophers feel that they know what the inescapable kind of *wrongness* might be? One factor may be the idea of a 'moral requirement' such that there can be one 'requirement,' as Williams says, on the agent to do A, and another requirement to do $B_{\sim A}$.[19] It is not, however, clear what is meant by 'a requirement' here. One wonders whether there may not be echoes here of the "divine law" conception which Anscombe has diagnosed in the contemporary moral philosophy that officially repudiates it?[20] We do, of course, understand the idea of a moral requirement in other contexts, as we might have said earlier that in cases where someone's claims have been overridden by a conflicting obligation restitution is "sometimes required." But here 'requirement' is explained in terms of a newly generated obligation, and so of wrongness of the common or garden kind, whereas 'requirement' itself needs explanation if it is to help with the special sense of 'wrong' that we are supposed to understand for cases of 'tragic choice.' Another contributing factor may be a certain schizophrenic tendency among moral philosophers who cannot bring themselves to say that there is any kind of action wrong in any circumstances whatsoever, and who yet feel that certain actions – say torture – cannot ever be justified. For it is exactly the 'tragic choice' situation in which they feel they must say that such actions would be justified, and yet feel that they must always be wrong. To this one would reply that either the "absolutism" of Aristotle, Aquinas, and Anscombe is (as I believe) correct or it is not. If it is correct and applicable to torture, then each and every torturer acts wrongly. Nor should it be supposed that two absolute moral prohibitions, which will relate to intentional actions, could conflict. For however terrible the inescapable choices that people have to make, they will never be between two intentional actions, such as for example, torturing X and torturing Y, but only between torturing X and *not preventing* another from torturing Y or from bringing about some other horrible result.[21] If, on the other hand, absolutism is not correct, then there could at least be hypothetical situations in which torturing is not wrong, and those who reject absolutism will have to bite the bullet about this.

However good these explanations may be of why *wrong if you do, wrong if you don't* has seemed intelligible, they bring us no nearer to actual intelligibility. What gets in the way is the fact that 'wrong' as understood in moral contexts applies to actions that count against a person's *goodness*: the goodness spoken of in the serious, nonironical, designation of an individual as one of the great and the good. So if we want to accept *wrong if you do, wrong if you don't* as an intelligible possibility, and still keep this negative relation between doing what is wrong and personal goodness, we have to revamp the latter notion to make that goodness vulnerable to the 'taint' of involvement in a horrifying, humiliating or tragic situation, or to the hatred of the gods: and then say that a choice that involves such *badness* is 'wrong.' I do not see how, ahead of having made these or some other shifts of meaning, we can know how to interpret the idea of a situation in which someone will necessarily be wrong whatever he does.[22]

NOTES

1. *Journal of Philosophy* 80, no. 7 (July 1983): 379–98.
2. *Proceedings of the Aristotelian Society* 29 (1965) and 30 (1966). Both articles are reprinted in Bernard Williams, *Problems of the Self* (Cambridge: Cambridge University Press, 1973), pp. 166–186 and 187–206.
3. "Moral Dilemmas and Consistency," *Journal of Philosophy* 77, no. 3 (March 1980): 121–36.
4. Bernard Williams, "Conflicts of Values," in *Moral Luck* (Cambridge: Cambridge University Press, 1981), pp. 71–82.
5. Marcus, "Moral Dilemmas and Consistency," p. 130.
6. John Searle, "*Prima facie* Obligations," in Z. van Straaten (ed.), *Philosophical Subjects: Essays Presented to P. F. Strawson* (Oxford: Clarendon Press, 1980), pp. 238–59.
7. Derek Parfit has pointed out to me that the testimony of others can be evidence of reasons for acting without itself being such a reason.
8. Williams, "Ethical Consistency," *Problems of the Self*, p. 180. See also Marcus, "Moral Dilemmas and Consistency," p. 134.
9. Susan Hurley, *Natural Reasons* (Oxford: Oxford University Press, 1989), chap. 7, sec. 3.
10. G. E. M. Anscombe, *Intention* (Oxford: Basil Blackwell, 1957), sec. 37.
11. For Donald Davidson's terminology see "Actions, Reasons, and Causes" in Davidson, *Actions and Events* (Oxford: Clarendon Press, 1980), p. 3.
12. I once read of a burglar who sat down to watch television in the house he was robbing, and got caught. He suffered from a lack of rationality that was additional to any that belonged to the project as a whole.
13. See, e.g., "Conflicts of Values" in Williams, *Moral Luck*, p. 76.
14. See Isaiah Berlin, "Political Ideas in the Twentieth Century," sec. III, in *Four Essays on Liberty* (Oxford: Oxford University Press, 1969) and the introduction to that volume.
15. We understand very well what it means when someone says on his death bed that he wasted his life on trivial things: on "things that didn't matter." Also when Berlin writes, again quite naturally, of "the best things in life."
16. See, e.g., Williams, "Ethical Consistency," *Problems of the Self*, pp. 172 ff.
17. See Williams, "Politics and Moral Character," *Moral Luck*, pp. 54–70; and Stuart Hampshire, "Public and Private Morality" in Hampshire (ed.), *Public and Private Morality* (Cambridge: Cambridge University Press, 1978), pp. 23–54.
18. Williams himself sometimes says that where an overriding obligation justifies the harm inflicted on someone he has the right to complain but is not actually

wronged. See "Politics and Moral Character" in Hampshire, *Public and Private Morality*, p. 61 ff.

19. Williams, "Conflicts of Values," *Moral Luck*, pp. 74–5.
20. G. E. M. Anscombe, "Modern Moral Philosophy" in Anscombe, *Collected Philosophical Papers* (Minneapolis: University of Minnesota Press, 1981), vol. III, pp. 26–42.
21. See ibid., pp. 34 ff.

9

Perspectival Guilt

PATRICIA S. GREENSPAN

One of the distinctive features of Ruth Marcus's work on moral dilemmas is its treatment of the question of guilt.[1] Marcus makes room both for appropriate guilt feelings (pp. 196–9) and for real or objective guilt (p. 194) on the part of the agent who fails, as he must, to satisfy one of the oughts in conflict in a dilemma. Her view rests on the assignment of responsibility to the agent for his choice of a specific option: Though he cannot avoid doing something wrong, since his choice situation offers him only wrong alternatives, he *can* avoid the particular wrong he chooses to commit; and this is enough to make sense of guilt. By contrast, Bernard Williams avoids talk of guilt feelings in favor of regret; and Bas van Fraassen, who does accept guilt feelings as appropriate, accepts along with them the doctrine of original sin as a way of allowing for real or objective guilt without responsibility.[2] In this paper I want to begin to provide my own line of defense for a view more like Marcus's, which I take to be derivable from a subtler account of the justification of emotions than that assumed by Williams and van Fraassen.

It was Marcus's work that led to my own sustained interest in dilemmas, and I shall make use of one of her comments at a crucial juncture in my argument, though an important modification of her position will also emerge as I proceed. Initially, however, I shall be moving rather far from the subjects she deals with herself to a discussion of emotional appropriateness – first in abstract terms (Section 1), then applied to the question of guilt and eventually to some disputed cases (Section 2). My treatment of cases will only gradually come into connection with dilemmas. What is at issue more generally is a range of cases of emotions apparently at variance with the agent's beliefs

The first draft of this paper was written while I was a fellow of the National Humanities Center during 1990–1, with funding provided by the National Endowment for the Humanities and the University of Maryland. About half of the paper was read at a symposium in honor of Ruth Marcus at the 1992 Pacific Division meetings of the American Philosophical Association. The bulk of the paper appears in chapter 5, sections 2–3, of my *Practical Guilt: Moral Dilemmas, Emotions, and Social Norms* (New York: Oxford University Press, 1994). I owe thanks to Kent Bach and to Walter Sinnott-Armstrong for comments on earlier versions.

about the surrounding situation. For guilt these include various problematic cases from the moral psychology literature such as guilt for the death of a child in an unavoidable car accident. I shall also consider collective guilt, survivor's guilt, and residual guilt for violating a childhood taboo. I shall then turn to cases of dilemma (Section 3), where I hope to sketch an argument *from* appropriate emotional guilt to a form of objective guilt that does not imply blameworthiness – the limitation representing my amendment to Marcus's views.

Let me begin by outlining some of the presuppositions of the general notion of emotional appropriateness that I shall be working from. The term "appropriate" is semitechnical: an artificially regimented refinement of our looser and more varied talk in common speech of particular emotions as reasonable or unreasonable in "backward-looking" terms – as fitting the agent's situation, accurately representing its salient or significant features insofar as he is aware of them. It is a notion of justification relative to reasons rather than a truth value of emotions; but it is meant to serve as their primary representational value in a sense that implies adequate performance of representational function.

Emotions are taken to have a representational function on the sort of account I favor just insofar as they are made out in terms of evaluations – meaning evaluative *thoughts,* possibly unasserted. I take these to be belieflike states or propositional attitudes with a content given by an evaluative judgment that might be thought of as corresponding to the emotion in question, though it need not always accompany the emotion.[3] The evaluative thoughts corresponding to emotions may not be connected logically to our other settled attitudes, that is, in the way that serves to pick out judgments as thoughts expressing *belief,* or attitudes of assertion. But they are held in mind in a more temporary way that allows them a significant influence on behavior. Normally they are objects of *attention,* though the account also allows for unconscious emotions as cases where the agent misidentifies the content of the evaluation.[4] What holds the content in mind in the relevant sense is emotional comfort or discomfort – something that can also be seen as having a representational function to the extent that it reflects the positive or negative aspect of an associated evaluation.

It is essential to this account that comfort and discomfort are taken as general intentional states, states of positive or negative affect directed toward evaluative propositions, rather than amounting merely to affective symptoms of emotional evaluation as in the usual list of sensory states characteristic of emotions. Thus, guilt amounts to discomfort *at or about* the thought of oneself as responsible for a wrong – not just to the thought *plus* an accompanying pang of discomfort. On the other hand, it is also important that in justifying guilt all we have to justify at least in the first instance is something on the order of a pang of guilt – a momentary state of discomfort tending to convey attention to its evaluative object in more or less immediate terms rather than the sort of settled behavioral tendency linked to belief.

Further, the justification of the emotion makes reference to its motivational function, with discomfort seen as a state that an agent would naturally want to get out of.[5] In justifying guilt what we are justifying is essentially an escape tendency: to change one's affective state by acting to falsify its evaluative content, or the self-attribution of responsibility for a wrong, seen as imposing a requirement of reparative action. That is, the point of the emotional reaction is to set up a need to act in certain ways that make up for and in some sense mitigate responsibility. But I take this to mean that the representational rationality of guilt (along with other emotions) will be properly influenced by instrumental considerations: The standards of emotional appropriateness will be adjusted to reflect facts about the general practical adaptiveness of a given emotion tendency, or its value as a motivator. For guilt my account will have the effect of extending the emotion beyond the corresponding judgment of fault – and beyond other-directed emotional blame, as we shall see, for an asymmetrical justificatory treatment of guilt and personal anger.

1. Emotions and Cognitive "Fit"

The account just sketched might be said to make out emotions as "evaluative affects." Comfort or discomfort and the thought toward which it is directed are not really separable parts of emotion; but I still speak of them as affective and evaluative "components," worth distinguishing conceptually in order to concentrate on the latter in considering justificatory questions. My strategy for considering emotional appropriateness involves packing the qualitative content of an emotion into its evaluative component – including the positive or negative aspect of the emotion, even though this might seem to repeat information given by the affective component – in order to permit an analogy to warrant for the corresponding belief.[6] The central aim of this approach, though, is to bring into sharper relief some of the limitations on the analogy. Let me try to capture these now at least roughly by appeal to a distinction in current philosophy of mind between "directions of fit."

The distinction arises in Searle's account of intentionality as a point of comparison between beliefs and desires (along with analogous speech acts) that applies only in a derivative sense to emotions.[7] A belief exhibits "mind-to-world" fit insofar as it is supposed to correspond to the world. By contrast, the point of a desire is to get the world to conform to *it*; its direction of fit is "world-to-mind." Thus, in the event of a disparity a belief counts as deficient in fulfilling its function, or in the usual terms false, whereas the onus normally falls on the world (including the agent) for failing to satisfy a desire. But emotions do not seem to fall neatly into either category.

Emotions are assessable in rational terms as appropriate or inappropriate – what I take to provide the analogy to belief warrant – but these terms normally build in reference to the adequacy of the reasons for an emotion, not just the sort of fit to the world that may be accidental. We may think of such reasons as facts about the agent, though, and hence as included in "the world" – at least if we take them as limited to noninstrumental or "backward-looking" reasons, assumed to characterize things as they are independently of the mental state in question, to allow for our initial contrast between states of belief and desire. A desire is also supposed to fit instrumental or "forward-looking" reasons, that is, insofar as these count as reasons for thinking its satisfaction good; but this is not fitness to "the world" in the sense that applies to beliefs. However, emotional appropriateness still resists characterization in these terms. Even assuming that it is *rational* appropriateness that is in question – as distinct from the various moral and quasi-moral or social versions of the notion that one might appeal to – it may not be clear that an emotion that fails to fit the world is rationally deficient.

Our assessment of states of sadness, for instance – to start with a simpler example than guilt – will not be based solely on whether they correspond to the facts of the case as measured by a single evaluative standard. The further fact that sadness tends to inhibit corrective action makes it a sometimes unhelpful response to misfortune. For practical purposes, then, we may be justified in withholding the emotional response in circumstances that warrant it. But this is also built into our view of things from a representational standpoint: Forgoing sadness is taken as rationally appropriate in a case where sadness would be too – for a basic initial contrast with the assessment of belief warrant.

On certain assumptions the same may be said of substituting a contrary emotion such as joy. Just to be happy about the same misfortunes described in the same way would of course be unreasonable; a certain caricature of an Eastern mystic might be able to manage it, but he would presumably be sacrificing reasonableness to something else. What is not so unreasonable, or unusual, though it also requires some mental self-trickery, is to focus attention instead on a different perceptual "slice" or cross-section of the information available in order to generate a positive emotion with preferable motivational effects. Insofar as the feeling thus generated involves attention to a thought that is reasonably picked out from the background of evidence for evaluation, the emotion will count as appropriate. But the point is that this justification rests on the assessment of a mental *act* – an act of attention – and hence is quite properly influenced by practical considerations.

What emotions are supposed to represent on this account is the importance of certain pieces of information – reasons of a certain sort, including reasons for attention, rather than directly "the facts" in some distinct sense. But attention is another element of cognition besides belief and is also subject to a kind of evidential assessment. Fitness to the salient facts is enough to warrant even a maladaptive emotion. Accordingly we may say that an appropriate evaluative object of affect – what my account makes out as the content of an emotion – amounts to one that on a reasonable view of things stands out against the background of evidence. Gestalt shifts are possible – there may be more than one reasonable view of things – so the notion is a tolerant one. It also has to be left somewhat fuzzy to correspond at least roughly to our intuitive judgments: What constitutes a reasonable subset of the evidence to pick out for attention – a reasonable perceptual "slice," as I put it – will often be a matter of debate. It is subject to moral and social norms in a way that allows for some degree of relativity to time and local convention. Consider, for instance, how we would classify sadness in the middle of a happy event at the thought of man's ultimate death and decay. The thought makes sense in a certain kind of religious culture but under other circumstances might be a matter for psychiatry.

The variability of the cases I have cited should indicate that we cannot expect an account of emotional appropriateness to provide firm answers to the many questions such cases raise. What I hope to provide is just an explanation of some of the relevant questions. First of all, our reason for picking out "slices" of the evidence for purposes of emotional reaction (and hence for its assessment) has to do with the function of emotions in directing attention: They serve to hold in mind reasons of potentially immediate or isolated significance – by contrast with evaluative *belief*, which essentially involves putting the same sorts of propositions into storage, connecting them with our other beliefs as settled response tendencies. As I noted earlier, this means that we need only justify something short-term in justifying an emotion – at any rate, if we limit ourselves to its basic qualitative justification, ignoring many important questions of degree.

The same point also seems to help explain why warrant for emotion and belief will be differently affected by the practical adaptiveness or instrumental value of a given evaluative thought. One should expect emotion with its basis in attention to be sensitive, most notably, to facts about the usefulness of a thought in motivating action – though as a quick response in many cases it cannot rest on any sort of calculation of consequences in the case at hand but at most can come to register the general value of the kind of thought in question. Belief, on the other hand, need not be borne in mind; and the results of including a certain evaluation among the objects of belief will be more unpredictable insofar as they are both less present to consciousness in standard cases and longer term. At any rate, as a normative matter, tied to our acceptance of representation as its defining function, we exclude considerations of adaptiveness from our weighing of the evidence for belief. This amounts to assigning belief its straightforward mind-to-world direction of fit. My alternative suggestion for emotions might be summed up as a claim that emotions exhibit *variable* mind-to-world fit to the extent that their representational function incorporates reference to forward-looking practical reasons along with the sorts of backward-looking reasons that constitute evidence. The standards of evidence are raised or lowered, tightened or relaxed, that is, in light of facts about the general practical adaptiveness of a given response tendency.

The resulting picture of emotional justification ties emotions to something more like perceptual standpoints than a unified conception of "the world." The range of allowable standpoints will be limited by facts of the sort that properly govern objects of attention – facts about the general significance of certain pieces of information – but since these are evaluative, they presumably do not themselves fit into the world in the sense intended. Even if one attempted to include them, however – as a kind of "atmosphere" to the world, say – the point for our purposes is that one would have to allow for multiple worlds (or a world with multiple overlapping atmospheres) to capture emotional appropriateness.

The notion of appropriateness that this account yields may be thought of as "perspectival" insofar as it allows for the assessment of emotions in relation to particular subsets or perceptual slices of the total body of evidence as well as in relation to the evidence overall. It rests on the general point that what we are justifying in justifying an emotion is essentially an affectively mediated tendency to direct attention rather than the sort of settled response tendency that is at issue in assessments of belief. The notion allows for appropriate emotion in cases of what I call "snap" evaluation, where we jump to an emotional conclusion before all the facts are in, as well as in other "parti-resultant" cases, with emotional deviation from warranted belief assumed to reflect the general significance of some part or aspect of the evidence. For guilt, I now want to say, the relevant aspect may amount to something like one's own involvement in a violation of the moral norms, even if the involvement falls short of moral responsibility – and even if the violation is not quite a wrong act – in contrast to the judgment that one *is* guilty.

My thought is that in general terms the subset of the evidence for the corresponding judgment that one responds to in these cases is important enough to justify at least an initial outlay of attention. This is what it means to be "sensitive to" the more ramified set of considerations that would ground the judgment; and there are further reasons in the case of guilt for thinking that *uncomfortable* attention might be warranted in particular as a way of controlling future behavior – or of making up for past behavior, if only by subjecting oneself to guilt. Guilt seems to be morally useful to us as a motivator, in fact, largely in anticipatory or symbolic forms that depend on perspectival appropriateness. In those cases where the emotion is warranted by serious past wrongdoing of one's own, adequate reparation is often just not possible. For that matter, it is questionable how often the emotion is felt in such cases – apart from those where the agent unwittingly failed to *see* things in the requisite light at the time of action, to appreciate the wrong done or his own involvement, and hence might seem to lack the fullest sort of moral responsibility. The point applies to cases of collective responsibility, where the main moral failing on the part of many agents is likely to be just the sort of failure to pay attention that emotional sensitivity on the account I have outlined is meant to protect against. It does so, if I am right, primarily by focusing attention on something short of adequate evidence for the corresponding belief.

2. Detaching Guilt from Blame

I now want to use the notion of perspectival appropriateness to explain why guilt should be appropriate in cases where blame – personal anger, its third-person counterpart in emotional terms – is not.[8] Guilt functions as a general way of keeping oneself alert to significant subsets of the available evidence, in the first instance for their potential bearing on future moral action. We can therefore justify the emotion as a reaction to subsets of the evidence that do not similarly warrant personal anger. It would be unjust, for instance, to blame someone personally in a case of collective guilt in which the agent himself did nothing wrong. The perceptual slice that *he* might reasonably focus on – the thought of the deed as chargeable to a group in which he claims membership – does not count as adequate reason for someone else to inflict emotional punishment on him. The aim of keeping him on his toes, morally speaking, would not justify the kind of treatment that one might reasonably accord oneself in a similar case – or demand of *him*. The reason is partly just that one *is* oneself, with more extensive sway over what one may do to oneself in emotional (or other) terms.[9] But further, guilt serves a positive function *for* oneself insofar as self-subjection to the emotion counts as a kind of self-cleansing ritual: a way of clearing oneself of involvement in wrong by emotional self-punishment. One would not similarly be benefiting another person, helping to clear *him* of involvement, by blaming him for the wrongs done by others in a group to which he belongs. To do so would be to indulge in a mental variant of scapegoating that subjects him to undeserved punishment on others' behalf.

This brief argument rests on making out emotion as essentially evaluative, with negative personal evaluations seen as unpleasant for their objects, even if their objects do not undergo the associated element of unpleasant affect.

The argument does *not* rest, however, on equating emotional blame with blaming others overtly: There are of course all sorts of moral or social reasons for restraining oneself from overt acts of blame; but even if we limit ourselves to mental acts, I take it that blame requires a stronger justification than guilt. Besides the sort of moral reason just given, the asymmetry turns on the wider set of motivationally adequate emotional alternatives to blame. In addition to the availability of a collective object of blame we ought to take note of a structural difference between blame and guilt that explains why I specify *personal* anger as the third-person counterpart of guilt corresponding to blame. For anger also has an *im*personal variant, frustration, whereas guilt is by definition directed toward the self.[10] But in combination with my perspectival account of appropriateness the availability of a less personally "punishing" alternative for anger means that the standards for justifying its personal variant will be higher.

Frustration counts as a "deficient" instance of anger, as I put it – in contrast to the full-fledged cases that represent the most fully elaborated instances of the emotion in adult life. So there is something else to turn to when the evidence is not adequate for blame. There may also be deficient instances of guilt – passive cases, say, without the characteristic desire for reparation – but my point here is just that these will not be deficient in quite the same way as impersonal anger. They will not be directed toward something more diffuse than their defining personal object, the self; otherwise, they would amount to something other than guilt, perhaps frustration or perhaps a form of sadness with an indefinite object (as on some accounts of depression). For anger, by contrast, generalized frustration, directed toward no one in particular, provides us with an emotional option to perspectival personal anger, anger that targets its object on an inadequate evidential basis for the corresponding judgment of fault. So the standards for imposing blame will not be relaxed as they are for guilt (along with other emotions), assuming that the general practical function of anger would be served well enough by frustration.

There are, of course, cases in which emotional blame without adequate evidence for a judgment of fault is of psychological or other benefit to the agent; cases of scapegoating might be thought to be chief among them. However, I take it that this does not affect the justification of personal anger in its representational function; we still speak of anger as unwarranted in such cases. The sorts of facts that affect our weighting of backward-looking reasons for the emotion have to do with its general adaptiveness – in the first instance, its energizing function in motivating corrective action. But I take frustration to involve a generalized urge to lash out, corresponding to its diffuse object, that should serve this end as well as personal anger in cases where the evidence does not pick out a specific object. In any case, it is not just benefit to the agent that is in question in considering the general adaptiveness of an emotion. Insofar as personal anger has a role to play as a *moral* corrective, by inflicting emotional punishment for injustice, any instrumental benefits of scapegoating would be undermined by the fact that it punishes an innocent party.

Guilt seems to be different, assuming that we have a right to scapegoat ourselves – to sacrifice ourselves to some aim of our own – at any rate, up to a point. The emotion has a positive function in defense of moral self-esteem that would not be adequately served by other-directed emotions like frustration – in general terms, that is to say: One might indeed recommend some other-directed substitute on many or even most occasions in the manner of various self-help manuals; but it is *general* adaptiveness that on my account lets us loosen our evidential demands on an emotion. And this is meant to be broadly construed, to extend to emotions that offer some unique benefits to the agent or just to people in general, even if they also inflict harms, perhaps even greater harms. Remember that perspectival appropriateness is taken as compatible with the reasonableness of *suppressing* a given emotion.

Let us now attempt to see what happens to some cases of guilt when we apply the notion of perspectival appropriateness. First consider the cases of collective guilt that have surfaced at points in my argument. My account of appropriate emotion allows us to grant that no one should be blamed solely for the crimes of others – that it would be unreasonable to hold someone responsible in a backward-looking sense (as opposed to insisting that he *take* responsibility) for what his nation or other group may have done in his name.[11] But on the other hand, we may also say that he ought to hold *himself* responsible in the way that is evidenced by emotion. He may be blamed if he fails to manage this – at any rate, in some form; there are emotional alternatives to guilt, albeit more limited than those that apply to blame.

The fact that guilt is appropriate, in the sense of "rationally acceptable," will not be sufficient to mandate that reaction, as I have noted. It does come out as appropriate, though, if we suppose that the agent's involvement in the group that counts as the collective agent in the case naturally sets up enough of a link between his own agency and the act in question to justify directing momentary discomfort toward a thought of himself as responsible for it. The thought may just be brought to mind with a question mark and rejected as soon as he has a chance to reflect on it; but it is appropriate as an object of uneasy attention as long as there is reason for him to *raise the question* of his moral responsibility. A crime committed by a group in the remote past, say, and clearly excluded from its future behavior by changes in group character over time would not naturally raise the question of responsibility in this sense, though of course a link could be manufactured.

The questions relevant to deciding such issues are of course extremely complicated, and the relevant concepts properly fuzzy. I shall be content just to sketch their application to the range of cases at issue. Consider now the case where one feels guilty for having caused the death of a child in an unavoidable car accident. A central cause of guilt feelings in such cases, or those we are inclined to accept as appropriate, is just the agent's need to reassure himself that he is *not* guilty. That is, after the accident the driver of the car that killed the child would naturally feel impelled to run through what happened, considering vividly his own participation in it, if only to be sure that he did not do anything to cause it. He would do so by confronting himself with various particular subsets of the total body of relevant evidence that considered in themselves would tend to suggest that he *is* guilty – and to call up feelings of guilt as they come to mind, given the responsiveness of emotions to partial evidence. In each case he would be able to answer any self-accusation by bringing to mind further evidence; and this would presumably allay guilt, though it would not prevent the reaction. And it would not prevent the reaction from recurring, at any rate for some time after the accident: A morally sensitive person in a case of such moral consequence could be expected to review the facts repeatedly and in rather exacting detail, focusing first on the memory of this or that act of his that might be picked out as a cause of the accident, then searching again for the reasons for thinking its results could not have been foreseen.

The need to *clear* oneself of real or objective guilt can in this way generate guilt feelings. Where the need is real, though, in the sense of being supported by more than the facts of the agent's individual psychology, the feelings may well come out as appropriate. It is not irrelevant to the case under discussion that the victim is a child, someone one is supposed to take special pains to look after, so that it is reasonable to assume more than the normal amount of responsibility even after the fact, and blame cannot easily be shifted to another agent. The agent reviewing his contribution to the case may focus, say, on the fact that he turned the car sharply in a certain direction – then remind himself that it was too dark to see the child – or he may focus on things he could have done differently whose connection to the case is more remote. Going out at all that night or going out with a trivial purpose might well come up as reasons for self-reproach (not just regret), at least until the facts of the case more or less settle in mind. My suggestion here is not that they should settle into a judgment of guilt but rather that emotional guilt properly *precedes* adequate evidence for such a judgment. Even when all the evidence is at hand, guilt will be appropriate in reaction to partial subsets of it whose role in producing the reaction serves a general practical purpose – in this case, as a way of assuming special responsibility for those unable to fend for themselves.

The appropriateness of the emotion depends, then, on the reasonableness in instrumental as well as representational terms of focusing attention on certain thoughts – including the sorts of perceptual memories one might well retain from a car accident and other bits of reflective information of a sort that can come to mind, be banished, and then come back again. This picture stands in contrast to the usual idea of even emotional guilt as imposed on the basis of an inner set of courtroom proceedings, deliberating to a final judgment. I would fill it out, moreover, with a parallel in mentalistic terms to the primitive notion of "tainting" in which the agent struggles to clear himself of a connection to some morally disturbing event by holding in mind the various thoughts that seem to link him to it and severing any putative link by means of emotional self-punishment.

What will be in question in disputed cases on the issue of appropriateness will normally be the general adaptiveness of guilt feelings in response to a subset of the overall evidence that constitutes prima facie evidence for a judgment of guilt. However, in cases of irrational guilt where the emotion is not based on a natural or reasonable perceptual slice of the available information I take it that we do not have even prima facie evidence for the judgment. This is how I would handle a variant of the accident case in which a memory of merely passive involvement – driving a car that was propelled into the child by another car that hit it – is "trumped up" into a recurring thought of oneself as somehow at fault, just on the basis of a neurotic habit of self-accusation.

Such cases are not so odd – nor are they limited to cognitively irrational agents – if one includes as genuine feelings of guilt the pangs and other momentary sensations that might be felt just in raising the question of warrant for the corresponding judgment in cases where responsibility has unclear boundaries. The heightened responsibility that adults assume on behalf of children yields a fund of familiar examples; but one might also consider, for instance, the feelings that arise as one asks whether damage to a borrowed book or a sublet apartment was something one ought to have foreseen and avoided. In some circumstances or states of mind – where one expects to be blamed unreasonably by others, say – it would not be unnatural to pay at least momentary attention even to perverse subsets of the evidence, if only to rehearse the arguments against according them prima facie weight. In general, it is important to note that not just any emotional response that might be made out as normal will come out as appropriate on the perspectival account, despite its appeal in standard circumstances to the notion of a natural object of attention.[12]

Some explanation of irrational guilt along the lines just sketched may thus be applied to more ordinary cases of guilt without fault from the literature. Let us see how it fits two particularly problematic sorts of cases. Consider first "taboo cases": cases of residual guilt for violating prohibitions from childhood or the surrounding society that the agent currently rejects. These resemble the original accident case in not involving self-attributed fault (meaning responsibility for a wrong), though unlike the variant case they do involve agency. But here the stress is on the denial of "wrong" rather than "responsibility." Gabrielle Taylor's inclusion of taboo cases as cases of genuine guilt (in contrast to Rawls's view) by appeal to conventionally accepted authority for the rule in question would seem to make them appropriate cases too – assuming no mistake about the conventions.[13] One might indeed provoke a pang of guilt just by remembering or anticipating blame on the part of others; so the response is in that sense natural. But I would expect us to reject as inappropriate a case in which the conventionally accepted rule in question just sets up an artificial way of sorting moral experience.

Proscribing something purely as a matter of convention, that is, without reference to human harms and benefits or other reasonable objects of moral attention (if there are any), does not support the appeal to general adaptiveness that results in a more relaxed standard of evidence for guilt than for blame. The general moral significance of emotion in a broader sense – as a way of enforcing conformity to whatever rule is in place – would seem to be relevant only where moral considerations do favor adherence to the particular rule in question. So "mere" taboos in the sense of purely conventional prohibitions whose validity one rejects will not yield even prima facie support for guilt feelings.

We have to say more, though, to handle taboos that may not be purely conventional but rather at least arguably have some basis – inadequate from the standpoint of moral justification – in a natural human response. Examples include fear of the stranger as a possible source of racial taboos and the feelings of hostility used to punish deviation from standard sex roles and preferences. On the assumption that a certain behavioral norm ought to be modified, the feelings that serve to enforce it need not come out as appropriate just because it is in fact widely accepted. However, rather than denying that guilt in such cases is based on a *natural* way of sorting information we can deny the adaptiveness in general social terms of allowing ourselves to react to a certain portion of the evidence in isolation.

On the other hand, there are rules on the order of the ban on incest that are commonly called taboos but are thought of as essential to the foundations of human society, albeit with details subject to convention. Here we may be willing to accept guilt as appropriate even in a case like that of Oedipus, where the violation is involuntary and hence does not merit blame. But this sort of extension of appropriate guilt beyond blame has to be limited to that phase of emotional justification that appeals to the special practical reasons for attention to a certain subset of the evidence for the relevant evaluation. The subset itself must be independently accepted as a natural or reasonable object of attention – on grounds I shall not attempt to pin down here – to avoid the kind of practical gerrymandering that would simply manufacture objects of guilt.

If we now turn to survivor's guilt and similar cases of guilt without agency – cases that might seem to be distinct in kind, involving the thought of separation from a group rather than of moral responsibility for a wrong – we can ask whether there is a way of counting them as appropriate without multiplying senses of guilt. Herbert Morris's defense of *nonmoral* guilt as a distinct notion is meant to allow for the view of basically rational survivors that they really *are* guilty.[14] In fact, many of the moral cases just discussed typically involve a component of survivor's guilt: The justification of collective guilt commonly appeals to benefits received as a result of the harms done to others by one's group; and even in the second version of the accident case guilt might be defended simply for surviving the child that dies. I would make out such feelings as involving a thought of moral responsibility but one that is not warranted by the evidence as a whole. Rather, on the perspectival account the justification for holding it in mind depends on the practical value of guilt as an identificatory mechanism – its role in promoting the sense of moral solidarity that Morris describes.

This is to say that guilt in the relevant cases functions as a way for the agent to clear himself via emotional self-punishment of any suggestion of benefit from being favored above the other members of a group with which he identifies. It expresses the fact that he does identify with the group by bringing him down to its level in emotional terms. But on the perspectival account his self-attribution of responsibility also has to rest on a natural way of sorting the available information. I think we can say that it does in serious cases of survivor's guilt – as opposed to just any case of guilt for the ills of the world – insofar as the thought arises naturally as a question in others' minds where people are genuinely "in the same boat" in some respect or other.[15] These are cases in which it is natural to resent inequality.

However, the account does not accept others' personal resentment of the agent as warranted in such cases. It matters to the justification of survivor's guilt that it is self-inflicted as a way of expressing identificatory commitments. Others are in no position to decide for the agent what these are to be; the most they can reasonably do is to demand that *he* decide a certain way as a condition of group acceptance. But even this does not pick out guilt in particular: The emotion comes out as optional, morally speaking as well as rationally, to the extent that other feelings such as those involved in humility represent alternatives for accomplishing the sort of leveling in question.

The cases I have brought up so far give us the beginnings of a continuum of cases on the question of grounds for guilt short of belief – and the beginnings of an argument from appropriate guilt *feelings* to a kind of objective guilt suitable for dilemmas. What varies in the cases is the strength of the reason we have for attention to the propositional object of discomfort involved in guilt feelings: the thought that the agent is responsible for a wrong. In the original version of the accident case but not the variant case or the typical taboo cases, we take it to be reasonable as well as understandable for someone in the agent's position at least to raise the question of guilt; but appropriate cases of collective guilt and survivor's guilt seem to involve something stronger: Unlike either of the accident cases they are cases in which the question of responsibility also naturally arises in *other* people's minds. There is a social as well as a nonidiosyncratic psychological basis for the emotion, that is – a basis for claiming that the agent really is tainted by what his society has done in his name, say; or that there is a real need to make up for inequality of the sort that undermines group solidarity. Other sorts of cases that Morris emphasizes such as guilt for thoughts seem to me to vary with the particular circumstances in respect to where they fall on the continuum. But I now want to extend the continuum in the direction of greater strength of reasons for attention–to cases where there is reason not just for raising the question of responsibility but also for reaching a positive conclusion.

3. Objective Guilt and Wrong

Moral dilemmas seem to be cases of appropriate moral guilt without full moral responsibility. They are cases in which the agent cannot avoid doing something wrong – even if he *can* avoid the particular wrong he chooses to commit – since his choice situation offers him only wrong alternatives. So the cases do involve agency – they even require specifically moral agency – but they seem to count as cases of guilt without fault, versions of the original accident case but distinguished by the agent's appeal to moral reasons for what he does. I take them to be cases in which each alternative is prohibited on serious enough grounds to make it come out as wrong ''all things considered'' despite the lack of permissible alternatives to it.[16] With the notion of perspectival appropriateness on hand and defended in intuitive terms for nondilemmatic guilt without fault, we can see dilemmas as cases in which emotional or subjective guilt is appropriate for either alternative. The thought is that the reasons for prohibiting a given alternative form a morally significant subset of the information bearing on a dilemma: a natural object of attention that is important enough in general moral terms to warrant an emotional reaction even without being balanced against countervailing reasons.

Indeed, we might be tempted to define dilemmas in subjective terms as cases in which either alternative warrants guilt. But despite my dismissal of rational guilt in taboo cases my notion of perspectival appropriateness could conceivably allow for cases of guilt without *wrong*. Perhaps guilt is sometimes adequately warranted in light of its general practical function in a case of merely prima facie evidence for counting some act as wrong, that is, as well as for holding the agent responsible for it. But in that case what the agent does in a dilemma might not really count as wrong all things considered under the circumstances.

Thus, we might say that the ban on parricide violated by Agamemnon's sacrifice of Iphigenia has enormous general moral importance that is registered in emotion but is not really in force as a guide to action in the particular situation – assuming that it is even *more* important to ensure the success of the Greek expedition.[17] Or in a ''balanced'' case like the one drawn from Sartre where the agent faces a choice between abandoning his dependent mother and failing in his duty to join the French resistance we might say the same for a prohibition of either action: Neither is more important, so both have merely prima facie status, and all the agent morally *must* do is choose – though guilt will be warranted whatever choice he makes.[18]

However, if we accept these cases as genuine dilemmas, at any rate as philosophers understand the term, I take it we have to say more than this. My own view is that the violation of prohibition in such cases is serious enough to count as wrong all things considered; it is not in the usual terms justified, then, at any rate fully, even if it is *excused* by the facts of the agent's choice-situation, so that he is not subject to blame.[19] But we can now see that the appropriateness of emotional guilt does not *imply* this view of the wrongs of the case. I shall go on to argue, however, that it does imply something stronger than merely prima facie wrong – something we can understand at least in rough terms by reference to the primitive notion of ''tainting.'' With some further backing, moreover, it can be made to yield a notion of objective guilt suitable for dilemmas.

My appeal to the notion of tainting in explaining the psychological function of guilt in my last section suggested a way of distinguishing among cases of guilt for the unavoidable. What we wanted there was a way of distinguishing between appropriate and inappropriate subjective guilt – on a notion of appropriateness that did not reduce to warrant for the corresponding judgment, of objective guilt or fault. But of those cases that come out as appropriate on the perspectival account it seems that some do and some do not involve what might be thought of as a real taint – a real need to dispel the *appearance* of fault, let us say – as a result of involvement in the act in question. I now want to suggest that this notion is sufficiently objective to allow for a reasonable interpretation of the comments of agents in survivor's guilt and related cases. It also will help us to see more clearly what is at issue in cases of dilemma.

I should first point out that the agent in the accident case is not tainted by involvement in a morally offensive act in the way that Agamemnon is, for instance. The case involves a need to clear oneself in response to a question that naturally arises about one's own responsibility for what happened; but as I understand it the case assumes that the question does not arise in other minds than one's own. It is taken as obvious, that is, that the agent in the case is not responsible for a wrong; indeed, as the case is set up, no wrong was done. What justifies attention to the question of responsibility and hence makes guilt an appropriate reaction is the adaptiveness on other occasions of the general emotional mechanism guilt brings into play. This is to say that there is no real or objective correlate of guilt in the particular case at hand, though the case has important motivational implications for other cases.

What I want to argue, though, is that more than this is involved in cases of dilemma – and even in the sorts of cases that Morris cites as examples of objective nonmoral guilt. As a first step in the argument we might say that these cases involve an *appearance* of fault – something that others besides the agent might be expected to react to, so that guilt functions essentially as a way of heading off others' blame. The agent's need to clear himself is based on something *intersubjectively* real, that is. I refer to it as a taint to mark what I take to be its nonaccidental resemblance to the primitive notion, which involved various punitive or cleansing rituals acted out even on inanimate objects used as instruments of wrongdoing.[20] But my point is not just about what we are inclined to *say* in cases of dilemma. For one thing, the verbal point does not really apply so readily to all such cases, as we can see by considering what we would say about the agent in the case drawn from Sartre. Failure in one's duties as a son or a citizen may not draw the kind of harsh reaction from others that leads us to speak of a taint as opposed to a serious moral shortcoming or other flaw. My point here is just that there is something, whatever it amounts to or should be called, that gives guilt more than individual psychological backing in such cases.

A taint in this sense amounts at least to an intersubjective tendency – to raise the question of responsibility – somewhat on the order of a *suspicion* of objective guilt in the semiobjective sense in which that term is sometimes used. To be tainted is to be "under a cloud" of suspicion, or reasonably subject to suspicion on the part of others – as opposed to just having reason oneself to raise the question of responsibility, which is all we need for appropriate subjective guilt as in the accident case. What we are doing here, one might say, is reinterpreting the primitive notion in psychological terms. An agent is *really* tainted on this account in cases where there is a real need to clear himself in the eyes of others – not just in any case where others would in fact be suspicious but only where in light of the facts or the evidence available (assumed to constitute a natural perceptual slice of the overall evidence) they would have real reason to be. At a minimum they have reason to demand an explanation. Emotional guilt is essentially a way of heading them off by raising the question oneself – and reacting to it, as an instance of a general response tendency learned for its motivational usefulness on other occasions that also serves on this occasion as a form of self-punishment symbolically distancing the agent from even nonculpable involvement in moral wrong.

To see how the notion of a "real taint" in the sense of intersubjectively warranted emotional guilt applies to cases – filling out the rational portion of the continuum of cases sketched in my preceding section – we might first consider variants of the accident case in which the agent does have "something to answer for" in others' eyes, even if he would in the end be declared blameless. One might suppose that the agent did something a bit irregular in his car upkeep that required explanation in light of what happened. Perhaps it was not really a factor in the accident; or perhaps its explanation would amount to an excuse. In any case, the very need to supply such a reason is enough to yield a limited objective basis for guilt on the account I have offered.

By contrast, I take it that on the usual interpretation of the accident case the need to supply a reason, to defend oneself against charges of objective guilt, is a matter of individual psychology, albeit by no means idiosyncratic. In general terms: A perceptual slice of the relevant information that counts as a natural object of attention from the standpoint of individual psychology may not have the same status from a social standpoint. In the original accident case others typically would dismiss the agent's guilt feelings as making no sense to them, just on the basis of what is immediately evident. But in other cases further questions may be in order – even if we sometimes suppress them in order to preserve social harmony or the agent's peace of mind. Despite its social reference, that is, the notion of a taint is not merely conventional: An agent may have "something to explain" independently of whether others in fact demand an explanation; and his real appearance of guilt in this sense will not depend on whether he actually appears guilty in others' eyes.

What should we say, then, about survivor's guilt and the other cases that Morris wants to treat as involving real or objective guilt? The word "taint" does not apply naturally to such cases; nor do they seem to raise a serious question of moral responsibility. Where they do involve some sort of wrong as in cases of being favored financially over others, it is taken to be something that is clearly not the agent's fault; what is objectionable to others about him is supposed to be just that he benefits from the situation, if only passively. However, I think we can still say that the agent in such a case has a need to prove something to others – to express his solidarity with them, bringing himself down to their level by way of the sort of symbolic leveling behavior I described previously. Guilt here may be seen as a way of keeping social frustration from focusing – as it naturally might – on him. What it serves to erase is a real appearance of indeed benefiting ("enjoying" benefits), or grounds for *envy*.

The notion of a taint does apply naturally to cases of collective guilt; and it provides a way of making out the real basis for collective guilt in application to individuals. In popular discussions of the question of German guilt in World War II, for instance, there sometimes seems to be a tendency to dismiss guilt *feelings* as unreasonable to expect from people born after the war – just on the grounds that it would indeed be unreasonable to hold them personally responsible for what happened. But assuming that there is a real basis for a judgment of *collective* guilt – in a sense, remember, that need not apply to individuals (as with corporations whose environmental sins cannot be charged to every employee – or in some cases, conceivably, to any employee in particular) – it makes sense to speak of innocent individuals as tainted by the acts of others. This does not mean that they bear some sort of permanent stain or are themselves morally flawed in some way but just that the question of moral responsibility naturally arises on the basis of their group affiliation – with emotional guilt supplying a kind of answer (though not the only possible answer) in practical terms.

The upshot of this discussion of cases – gradually narrowing down my initial cases of guilt without fault by appeal to the perspectival notion of appropriateness – is that there is sometimes a real basis for guilt feelings that falls short of the judgment that one *is* guilty. I have referred to this as a taint, offering an intersubjective reading of what I take to be a successor to the primitive notion of tainting. But we might just think of it instead as perspectival *guilt*. It amounts to something midway between guilt feelings and objective guilt in the usual sense, as implied by a judgment of fault; and it seems to be enough to explain those nondilemmatic cases of guilt without fault in which we would not be inclined to dismiss the agent's claim that he really *is* guilty. Instead of reconceiving the cases as involving guilt in some other, nonmoral sense we may reinterpret their claim to objective status in terms of intersubjective backing for guilt *feelings*.

Real taints in this intersubjective sense would seem to represent the strongest consequence for dilemmatic options that is derivable from an argument from moral feeling, an argument based solely on the appropriateness of guilt or some similar emotion in response to whatever the agent does. However, by adding in reference to the element of voluntary control at this point in the argument, I think we can see that dilemmas also involve something stronger, something closer to full-blown objective guilt, though still perspectival in a way that accords with my treatment of the grounds for the emotion. They involve a real *basis* for full-blown objective guilt – not just for raising the question of responsibility but indeed for a positive answer in the form of a judgment of fault, though the evidence for that conclusion is overturned by countervailing reasons. In itself, that is, the evidence is sufficient for the attribution of responsibility just insofar as the agent intentionally did something he knew to be wrong under the circumstances; but since they were circumstances where he *had* to, morally speaking, a broader view would show that he is not really at fault.

Thus, even long after Agamemnon's sacrifice of his daughter, the act requires explanation in moral terms. Objectively speaking rather than subjectively (if he feels nothing), there is a need for Agamemnon to satisfy himself and others that the act was morally necessary under the circumstances. But we also want to say more than this: that the act left moral traces of a sort that are summed up in the imagery of staining. It is a mark against him, a blot on his record of moral action and a threat to his character or virtue, unless the wrongness of the deed (from a limited but morally central standpoint) can somehow be ''put into perspective'' by rehearsing the explanation for it yet again.

The way dilemmas apparently differ from the cases of intersubjectively real taints just considered lies in something like the fact that Marcus notes in defense of dilemmatic guilt: They do involve responsibility for a *particular* wrong, insofar as the agent in them chooses a wrong option in the knowledge that it is wrong and carries through on the choice by voluntary action. Of course the choice is in some sense *in*voluntary or forced on him; he has no control over his range of options, in particular the fact that all of them are wrong. So he is not responsible for doing *something* wrong, or for wrong or wrongdoing as such.[21] But he is responsible for the particular wrong he does in a sense that exceeds mere involvement in moral wrong. We might say that his act is culpable as far as it goes – perspectively, in other words, or considered from a certain standpoint, a standpoint that is important enough to count as morally significant in itself.

This is more than a claim of *general* moral significance, of the sort that yields feelings of compunction in response to prima facie wrong. My suggestion in Agamemnon's case is that there are strong enough reasons against the sacrifice of his daughter to make it count as wrong all things considered *in* the circumstances in which it occurs even on the assumption that there are stronger reasons on the other side. That these reasons for counting the act as wrong also tend to undermine the agent's moral worth – or metaphorically, that the act stains the self – counts as an objective basis for felt guilt that still makes psychological reference but is not therefore merely intersubjective.

Marcus's own treatment of dilemmatic guilt does not seem to be limited to a first-person standpoint in the way I have in mind, though. She appeals, as I have just done, to the agent's voluntary control over what he does in support of the attribution of responsibility. But her comments do not distinguish between grounds for guilt on the part of the agent and grounds for blame, the corresponding other-directed emotion. Our intuitive view, I take it, is that blame in the sense of personal anger is not really warranted in such cases – at any rate, for what the agent does, as opposed to a failure to feel what he ought to feel about his action.

It may be reasonable to *shun* an agent tainted by a seriously wrong act – an act on the order of parricide, say, as in Agamemnon's case – or to feel various related emotions (possibly taken as variants of blame) such as personal aversion or horror. But if we really agree that he made the morally best choice open to him at every stage, it would not be appropriate, though it might be natural, to blame him in the sense that entails feeling the desire to punish him characteristic of anger. From a third-person standpoint on the perspectival account his act comes out as wrong but excused, even though in emotional terms he should not excuse *himself* and is therefore subject to blame for not feeling guilty. According to the argument I have outlined, this asymmetrical treatment of blame and guilt rests on the differential practical effects of reacting to a partial basis for assigning fault – assessed in light of the different emotional alternatives to that reaction – in one's own case versus that of another. On Marcus's unamended account, by contrast, an agent in a dilemma would come out as morally blameworthy for his action – for doing what morally speaking he had to do.

To complete the task of characterizing dilemmatic guilt perspectivally, let me now attempt to show how we might replace the metaphor of staining with appeal to *reasons*, of the sort that would tell us when a taint counts as real. As noted earlier, talk of taints is not quite applicable to cases like the one drawn from Sartre. On the other hand, it does apply to nondilemmatic cases of moral conflict. We speak of money from a bad source as tainted, for instance, even on the assumption that it would not be wrong to make use of it for a good end. Instead of speaking of taints, then, if we think of emotional guilt as a self-cleansing ritual in the way I have suggested, we may say that the agent in a dilemma has real grounds for it insofar as the ritual is really needed to cleanse him of responsibility for a wrong.

This means that the reasons supporting a negative self-evaluation, even if evidentially limited, are morally important enough in themselves to affect the assessment of the agent's moral worth unless checked – as they are when the agent subjects himself to guilt. To remove the metaphor of tainting, then, let us stipulate, first, that the reasons are significant enough for the evaluation to register in attention in the way that unpleasant affect allows it to do – thus warding off a more settled negative self-judgment. For dilemmas, however, we need a second stipulation, one that explains why the need for the self-cleansing ritual recurs: From the limited perspective of the act under consideration we do have enough reason for reaching a judgment of fault; the judgment is undercut only in light of the total body of evidence. So attention to the act (which is morally warranted) warrants emotional guilt. What we have in the role of objective guilt in dilemmas, then, is essentially an unsatisfiable need – a need to dispel the appearance of fault, based on a limited subset of the evidence that is important enough by itself to yield moral wrong.

NOTES

1. See Ruth Barcan Marcus, "Moral Dilemmas and Consistency," *Journal of Philosophy* 77 (1980):121–36; reprinted in C. W. Gowans, (ed.), *Moral Dilemmas* (New York: Oxford University, 1987), pp. 188–204. I shall give page references parenthetically in my text to the reprinted version in what follows.
2. See Bernard Williams, "Ethical Consistency," in Gowans, *Moral Dilemmas*, pp. 115–37, pp. 123–35, and Bas C. van Fraassen, "Values and the Heart's Command," ibid., pp. 138–53, p. 147. Cf. also Michael Walzer, "Political Action: The Problem of Dirty Hands," in M. Cohen et al., *War and Moral Responsibility* (Princeton, NJ: Princeton University Press, 1974), pp. 62–82, p. 68 (cf. pp. 71–4), for another defense of guilt in dilemmas but without the particular line of argument Marcus provides.
3. For my full argument on this point see *Emotions and Reasons: An Inquiry into Emotional Justification* (New York: Routledge, Chapman and Hall, 1988), esp. chap. 1, pp. 1–80; for the application to guilt cf. "Subjective Guilt and Responsibility," *Mind*, 101 (April 1992); 287–303, and *Practical Guilt*, chap. 5, sec. 1.
4. See *Emotions and Reasons*, pp. 25–30.
5. Cf. ibid., p. 31 (cf. pp. 153–62) and "Subjective Guilt and Responsibility," p. 9.
6. Cf. *Emotions and Reasons*, esp. chap. 4, pp. 83–107, for my extended inquiry into appropriateness, resulting in a nonquantitative criterion in terms of belief warrant (p. 106).
7. See John Searle, *Intentionality* (Cambridge: Cambridge University Press, 1983), esp. pp. 31–5; cf., e.g., p. 8 for his statement of the point as a claim that emotions have *no* direction of fit–meaning, I take it, emotions themselves as distinct from desires and beliefs. Searle himself sees emotions as combining elements of desire and belief, with an emphasis on desire (but as only partially characterized by his analysis; cf. p. 36), while making out appropriateness in terms of belief satisfaction (pp. 8–9). Note, however, that he uses the category of desire to capture evaluative content.
8. Cf. Allan Gibbard, *Wise Choices, Apt Feelings: A Theory of Normative Judgment* (Cambridge, Mass.: Harvard University Press, 1990), e.g., p. 139. In the ensuing argument I hope to show how the standards for guilt and anger might diverge compatibly with the value placed according to Gibbard on meshing guilt and anger (cf. ibid., p. 299). See Walter Sinnott-Armstrong, *Moral Dilemmas* (Oxford: Blackwell, 1988), p. 107, for the application to dilemmas of a similar claim of asymmetry between *remorse* and blame, though without the explanation I provide here; cf. also my "Moral Dilemmas and Guilt," *Philosophical Studies*, 43 (1983): 117–25.
9. Cf. my more extended comments on the self-versus-other asymmetry in application to guilt versus blame in *Practical Guilt*, esp. chap. 4. I here bypass in particular a detailed treatment of the contrast with other self-directed moral emotions, most notably *shame*.

10. Cf. *Emotions and Reasons*, pp. 48–55, and p. 188, n. 3.
11. Cf. my distinction between backward- and forward-looking notions of responsibility in "Unfreedom and Responsibility," in F. Schoeman (ed.), *Responsibility, Character, and the Emotions* (Cambridge: Cambridge University Press, 1987), pp. 63–80. One might be tempted to take blame as appropriate just in virtue of its usefulness in getting individual members to exert forward-looking responsibility and *reform* the relevant group. However, *general* adaptiveness would seem to be well enough served by blame imposed *for* failing to *take* responsibility–in that way and also in the emotional terms that I go on to justify. Something similar applies, moreover, to any independent symbolic function that might be assigned to blame, in expressing a strong stand against the group behavior in question.
12. On the other hand, the standards of appropriateness may incorporate facts about normal emotional *development* to the extent that these affect both the naturalness of an object of attention and the general adaptiveness of an emotional response to it. This point may afford a deeper explanation of Ronald De Sousa's notion of "paradigm scenarios" [see *The Rationality of Emotion* (Cambridge, Mass.: Bradford Books, 1987), e.g. pp. 181–4], as early stages in emotional learning that are given special stress in determining the rationality of an emotion in adult life. In view of facts about emotional "inertia," or the tendency of emotional responses to outlive changes in their corresponding judgments, we would expect some stages in childhood development to affect adult response patterns. Thus, for instance, it might be thought that anger in response to injuries to oneself, as the natural childhood strut of the sense of justice, is important enough in developmental terms to warrant blame on the part of the victim in cases of dilemma. But there is room for a good deal of empirical disagreement on these issues; it should be evident that my treatment of cases here rests on a heavy element of conjecture.
13. See Taylor, *Pride, Shame, and Guilt: Emotions of Self-Assessment* (Oxford: Clarendon Press, 1985), p. 86; cf. John Rawls, *A Theory of Justice* (Cambridge, Mass.: Harvard University Press, 1971), p. 482.
14. See "Nonmoral Guilt," pp. 220–40, in Schoeman, *Responsibility, Character, and the Emotions*, p. 221. Note that Morris also makes out collective guilt as nonmoral–a point I have questioned elsewhere; cf. "Subjective Guilt and Responsibility," pp. 10–11 ff. Cf. also Morris's somewhat less discriminate acceptance of "shared guilt" in *On Guilt and Innocence: Essays in Legal Philosophy and Moral Psychology* (Berkeley: University of California Press, 1976), pp. 111–38.
15. Cf. my "Guilt as an Identificatory Mechanism," *Pacific Philosophical Quarterly* 74 (March 1993): 46–59.
16. See *Practical Guilt*, chap. 2, sec. 1.
17. Cf. Aeschylus, "Agamemnon," in *The Oresteian Trilogy*, trans. Philip Vellacott (Harmondsworth, Middlesex, England: Penguin, 1959), pp. 41–100, pp. 48–51; for the use of this case as an example of moral dilemma see esp. Williams, "Ethical Consistency," p. 123; cf. Martha Nussbaum, *The Fragility of Goodness: Luck and Ethics in Greek Tragedy and Philosophy* (Cambridge: Cambridge University Press, 1986), pp. 25–82.
18. See Jean-Paul Sartre, *Existentialism and Humanism*, trans. P. Mairet (London, 1948), pp. 35–6.
19. Cf. Sinnott-Armstrong, *Moral Dilemmas*, p. 42; unlike Sinnott-Armstrong I take it that "I had no choice," even if meant as an appeal to *moral* necessity, counts as an excuse. For my distinction between "all things considered" and "overriding" cf. *Practical Guilt*, chap. 2, sec. 1.

20. Cf. George P. Fletcher, *Rethinking Criminal Law* (Boston: Little, Brown, 1978), pp. 343–4.
21. This point distinguishes dilemmas from the cases of responsibility for inevitable states of affairs cited in John Martin Fischer and Mark Ravizza, ''Responsibility and Inevitability,'' *Ethics* 101 (January 1991): 258–78, esp. pp. 258–64. Note that the agent in a dilemma *would* have done otherwise (avoided doing wrong) if he had been able to; cf. the contrasting cases cited in ibid., pp. 264–5.

10

Instigating the Unpredisposed: Bad Luck in Law and Life

JOEL FEINBERG

Causing Another Person's Voluntary Behavior

There are many different ways of getting a person to act as you want him to act. Aristotle quite plausibly argued that only some of these ways, those employing force or a certain kind of deception, provide the actor with an excuse for his action when what he did was wrong. Among the forceful techniques, we can list ordinary compulsion, in which the instigator, for example, pushes or kicks the other, or fastens a chain around him and drags him along the ground. More likely, the forceful instigator will threaten the second party with harm to himself or others to be inflicted if he fails to do what the threatening party wants him to do. This technique, which resorts to force only indirectly as a backup to a threat, is commonly called coercion or duress.

Similarly we can defeat a charge of wrongdoing by showing that another person got us to do the wrongful act by providing us with false information so that the act we believed we were doing, the act we intended to do, was not in fact the act we did. I may think that a nun in her habit is soliciting funds for a church charity (as she says), when in fact she is no nun at all, but a man masquerading as a nun in order to raise funds for himself. In that case, I can argue that even though what I did (put a dollar on his plate) was under one description voluntary, under another it was not. I did not voluntarily contribute money to the private bank account of a criminal, and therefore I cannot be held responsible after the fact, for any act under *that* description. Similar techniques, however, lose their exculpatory effect when the deceiver is a policeman baiting his hook to catch a criminal who has a known readiness to commit a certain kind of crime. If *B* offers to buy a packet of cocaine from *A*, apparently a willing and eager drug dealer but actually a police agent in civilian dress, *A* is deceived about no element of the crime he is bent on committing, but only about the identity of his customer and his chance of getting away with it. It is no excuse to the charge of illegal drug sales that the seller believed his customer was not a policeman.[1]

On the other hand, there are various other techniques of manipulation that are quite consistent with the manipulated party's responsibility. Some of these can be as effective even as compulsion and coercion in getting him to do what his manipulator wants him to do, and yet they are not themselves uses of force or coercive threats. *B* may goad or prod *A*, persistently coax or implore him, scold or berate him, importune or cajole him, all to a degree that amounts to pestering him. For convenience we can label this somewhat miscellaneous category, *goading* techniques. *B*'s conduct in all these examples falls short of using force or violence and makes no explicit threat of harm. If *A* lets it "get him" to do *B*'s bidding, he cannot escape at least shared responsibility for his actions, even though it is unlikely that he would ever have performed those actions if *B* had not got him to do so.

A second category of manipulative techniques that do not defeat the subsequent responsibility of the manipulated party we can label *luring* techniques. In these cases *B* exploits *A*'s weakness of will by providing some sort of enticement to the wrongful action he wishes him to perform. *A* is not forced so much as beguiled, attracted by a prospect rather than repelled by its alternatives. The hope of gain can be a powerful motivator indeed, but it is rarely if ever thought to be the kind of consideration that excuses wrongdoing. Violence and duress are often irresistible, but "irresistible temptations" are never allowed to exculpate. In any event, what is crucial for my purposes in this essay is that in all the types of induced actions with which we shall be concerned, whether the inducements be goads or lures, the initiative is wholly *B*'s, but the retrospective responsibility is at least partly *A*'s.

Entrapment in the Criminal Law

Police officers have frequently used *agents provocateurs* and similar techniques to "get people" to commit crimes so that they can be caught in the criminal act, arrested, convicted, and incarcerated, their dangerous propensities no longer likely to produce harm to the people who would otherwise be their victims. Sometimes the disguised police officer has little to do but make known to the criminal that he has an opportunity to commit the crime, and the eager criminal, already predisposed to take advantage of such chances, leaps to the bait. Thus, a secret agent may offer to purchase a prohibited drug from a suspected drug dealer who immediately agrees, asks a certain price, hands over the drug, and pockets the agent's cash. Even before the dealer can say, "Thank you, have a nice day," the agent identifies himself, announces that the vendor is under arrest and puts him in handcuffs. In other cases, however, the police plot is more complicated, and the police agent's role more active, so that there is doubt whether the party the police entrap is an unwary criminal or merely an "unwary innocent" tricked or lured into committing a crime he probably would not otherwise ever have done.

"Entrapment" then has become a derogatory term, referring to a governmental abuse of power, so that for a defendant in a criminal trial to prove police entrapment now is for him to use a defense that can get him off the hook of criminal liability. The classic definition comes from the first Supreme Court decision to recognize and formulate an entrapment defense, *United States v. Sorrells*[2] in 1930: "Entrapment is the conception and planning of an offense by an officer and his procurement of its commission by one who would not have perpetrated it except for the persuasion of the officer."[3] This defense is obviously not available to the drug dealer who quite willingly agrees to sell a prohibited drug to a secret agent. He was ready and willing to engage in that criminal behavior whenever he had the opportunity. Similarly, the defense is not available to the pursesnatcher who takes the purse of a secret "lady cop" who carries it strapped over her shoulder in the common manner, in a crowded department store, and makes no special effort to lure or encourage him. In other cases, however, it may seem clear that but for "repeated and persistent solicitation," the defendant would not have done anything illegal then and there, and probably would have spent the rest of his days as a respectable law-abiding citizen. That kind of defendant needs an entrapment defense, and gets one in virtue of the decision in *United States v. Sorrells*. The Supreme Court in that landmark case overturned Sorrells' conviction for the illegal sale of an alcoholic beverage during the Prohibition era.

Leo Katz tells the story:

> Sorrells, who lived in a small town in North Carolina, received a visit from several of his friends. They brought with them a stranger named Martin, who introduced himself as a furniture dealer from Charlotte, just passing through the town. When Sorrells discovered to his great pleasure that Martin and he shared a common background – both having served in the Thirtieth Division of the American Expeditionary Forces during World War I – the two became fast friends. Martin, Sorrells, and another friend of Sorrells . . . then began to reminisce about their war experience. After a little while Martin asked Sorrells for some liquor. Sorrells apologized because he didn't have any. Some more time passed, and the conversation was taking an increasingly nostalgic turn. Martin again asked Sorrells for some liquor. He explained that he was anxious to buy a half gallon for a friend back home. Sorrells replied that he did not "fool with whiskey." As time went on Martin reiterated his request for liquor some six or seven times. Finally Sorrells got up, excused himself for about twenty minutes, and returned with the requested half gallon. Martin gratefully accepted and paid him five dollars. Alas, Martin turned out to be a prohibition agent posing as a tourist. . . .[4]

The Supreme Court considered carefully just how Sorrells's conduct was instigated by the prohibition agent. The criminal act, it said, was "a creature of *his* [the agent's] purpose." Sorrells, it seemed clear, was both pestered by repeated and persistent solicitations, and lured by nostalgic sentiments of comradery, in the absence of either of which he would surely not have sold liquor to the agent. Perhaps the Court spoke a bit carelessly when it asserted that Sorrells had "*no* previous disposition" to do what he did, and that the agent "implanted in his mind" the disposition to commit the offense. Since Sorrells did what he did voluntarily, he must have had *some* prior disposition to act as he did in the sort of circumstances, rare and contrived as they were, that obtained. But *that* disposition – the tendency to act against one's own principles or interests when pestered, deceived, and lured in just the manner in which he *was* pestered, deceived, and lured by his instigators, should not have been inculpatory. The prohibition agent did not so much *implant* the disposition as (to switch to a better metaphor) *trigger*[5] the weak disposition that was already there.

Usually factors of two kinds, the police inducement and the defendant's predisposition, are debated at a criminal trial in which the entrapment defense has become the chief issue. It is normally up to the trial judge to decide whether the police inducement had been so inordinate that the trial must be terminated and the police reprimanded, no matter how telling the evidence of the defendant's readiness to commit the crime would have been had it been admissible. If the trial is permitted to proceed, then the defendant's prior readiness to commit the crime that was then encouraged by secret police inducers must be proved to the jury, along with the elements that define the crime.

Generally, the weaker the police inducement, the more "ready" or "predisposed" the defendant must have been to commit the crime anyway, even without the inducement. Similarly, the stronger the police inducement, the less likely it is that the defendant was prepared to commit the crime, and the less stringent the independent evidence must be that he was in fact so predisposed. In the limiting case where police agents are involved only minimally in the creation of the crime, say by merely providing the defendant with an opportunity, hardly any evidence at all is required of the prisoner's criminal predisposition. If as little influence as *that* could produce his immediate criminal response, he must have been predisposed. The case for the weakness of the inducement itself implies or produces the case for a strong predisposition, just as a case for the weakness of the predisposition would itself imply a case for a strong, possibly "inordinate," inducement. In such ways the arguments against the inordinate inducement and for the criminal predisposition become intertwined, each a part of the other.

Scams and Stings

The most publicized examples in recent years of prima facie improper governmental inducements and blameworthy predispositions have been elaborately designed Federal Bureau of Investigation scams by which the government casts its nets widely in the hopes of convicting large numbers of corrupt officials, including such powerful persons as U.S. congressmen. The most interesting aspect of these cases, I think, is that the entrapment argument employed by the defense did not prevent juries from convicting the celebrity defendants. Since it appeared, in the leading case[6] at least, that there was something approaching inordinate inducement from the FBI and only a very weak degree of readiness on the part of the defendant congressman, it may well have been the jury's view, and a plausible view it is, that the office of a congressman is so elevated and powerful that a congressman must be held to a higher standard of probity than that which is customarily applied to ordinary citizens. She must be more than relatively difficult to corrupt; she must be as close to absolutely incorruptible as human nature will permit.

This conception of the responsibility of congressman was shared by the law enforcement officers who were motivated by it to establish the so-called ABSCAM (short for "Arab Scam") program in 1981 to test the integrity of high government officials. Toward this end the FBI hired a professional con man with a long criminal record of such crimes as extortion and fraud, one Melvin Weinberg, who posed as an Arab sheik, caftan and all, and assumed an Arabic name. Leo Katz tells the story from that point:

> Weinberg let it be known that the sheiks were concerned that a change in government in their homeland might force them to emigrate to the United States. To insure against any problems they might have should the occasion arise, Weinberg indicated, the sheiks wanted to "sign up" as many congressmen and other public officials as possible. The congressman could expect, in return, sizable investments in their districts and payments of cash to them personally. The first of those baits was, of course, entirely legitimate; it was the second only that constituted a bribe.[7]

Weinberg's "prodding" of Representative Kelly, as Katz puts it, was at times "relentless" – similar to that of Mr. Martin in the *Sorrells* case, but more powerfully alluring and more persistently pursued. Weinberg, posing as an Arab sheik, made an appointment with Kelly. He found the congressman interested in the offered investment in his district (as indeed he should have been, according to Katz), but determined to reject the personal payoffs out of hand. He wanted "no part in that." Through a number of subsequent meetings, Kelly remained adamant. Then, like Sorrells (but from quite different motives) he gave in "when $25,000 was displayed for him spread out in packets of $100 bills. He stuffed the bills into his pocket and left."[8]

Some readers may feel no sympathy for Kelly whatever. But I find that my reaction is both confused and ambivalent. On the one hand, Kelly betrayed his trust as a high government official. He didn't *have* to give in to the temptation. Neither compulsive pressure nor coercive threat was used against him. Even the FBI deception was not of the sort that would diminish the voluntariness of his conduct. He didn't voluntarily walk into an FBI trap to be sure, but he did voluntarily commit what appeared even to him to be a crime. On the other hand, Kelly was induced to act as he did by police agents; his criminal conduct was their idea, and their initiative. Without their instigation, certainly Kelly would not have acted as he did then and there, and the probability of his *ever* acting in that fashion was very low. Kelly was a middle-aged man who had never even been *accused* of a crime. His character was thought by all who knew him to be impeccable. But for the deliberate trickery of the government in artificially creating his crime, his perfect record would probably have remained forever untarnished.

Perhaps a case can be made that Kelly has been treated unfairly. But there is something unseemly about *his* being the one to voice that grievance. Gerald Dworkin has a point when he says that the trapper should be treated as a "tester of virtue not a detector of crime,"[9] and even Kelly will have to admit that he has flunked the virtue test. There is no way in which his wrongful action can be justified or excused, wherever the original initiative for that action is located. The combination of these points seem to yield a moral paradox. Kelly was treated unjustly, as "a means only," yet his behavior, though initiated by the unjust parties, was nevertheless his own, somehow expressive of his own character, and therefore his own responsibility. After all, the government didn't force him to take the bribe money. Neither compulsion nor duress was used. Rather the instigators lured, goaded, coaxed, and implored; they urged and encouraged, and in the end, given their success, perhaps it can be said that they *seduced* him into doing it. But these verbs are not the names of legal defenses. The law rightly refuses to accept as an excuse the claim that another party made a criminal act seem so attractive an idea that he could not resist its appeal. It seems clear that the FBI in the Kelly case "got" their man to do what is criminal and to do it voluntarily, on his own responsibility.

Are We All Criminals – More or Less?

Why do we find the idea of entrapment so disturbing? Leo Katz suggests that our moral discomfort with entrapment stems from our belief or fear that there are criminal dispositions in all of us. For each forbidden act, we suspect that there is some set of circumstances, however rare and unlikely, in which you, the reader, would voluntarily commit that crime, and similarly for all the rest of us. It is more plausible, of course, to think of everyone as capable of committing *some* kinds of forbidden acts (e.g., overtime parking) but not others (e.g., first-degree murder). Nevertheless, one can learn some alarming facts about what one would be likely to do in certain extraordinary circumstances if one has a sufficiently resourceful imagination, or better still if one reads many books by imaginative and insightful novelists. I remember how shocked I was to learn that I could – indeed that I probably would – commit murder with a more or less clear conscience in circumstances like those that enveloped the protagonist in Ludwig Lewisohn's horrifying novel, *The Case of Mr. Crump*. That novel, written in 1926, is the story of a woman who manipulates a gifted and sensitive man, twenty years her junior, into a marriage, and then exploits his weakness and inexperience, as well as the stupidities of the old New York state divorce laws, to keep him in a hellish relationship that ruins his career, renders impossible his ambitions, and finally steals from him his last and only chance of love. In the end the man has no alternative but to destroy this ruthless virago. He does, and so probably would I if *mirabile dictu* I were in his shoes. In short, I have a disposition even now, as I write, to commit murder if I were in the extraordinary and highly improbable circumstances of the fictitious Mr. Crump. Fortunately, no one has anything to fear from me, given how rare and unlikely such circumstances are. It is important to note, however, that it is not so much the character of my dispositions themselves that give reassurance, as the improbability of the circumstances that could activate them, and the occurrence or nonoccurrence of such circumstances is, in a sense, a matter of luck, beyond my control.

Tests of Strength

It is the existence of a predisposition, according to the Supreme Court, that determines the guilt of an entrapped person when the entrapment is legitimate. This upsets and alarms people, as we have seen, when they come to suspect that all of us have predispositions to various crimes if only in the very dilute sense that there are wildly remote and unlikely circumstances in which, acting in character, we would perform a forbidden act. There cannot be much blameworthiness attached to *that* kind of disposition, surely. But dispositions differ in strength; one person may be *more disposed* to do something than another person. Correspondingly, we would expect the stronger dispositions to be the more blameworthy ones, other things being equal. We have already encountered some of the ways in which there can be a "more" or a "less" of human disposition along some relevant quantitative dimension. A disposition is expressed in the form of one or more hypothetical sentences predicting how the possessor of the disposition would act in various kinds of circumstances. Our first quantitative measure then might be to determine how likely or unlikely the precipitating circumstances are to occur. If the United States were to drift into a form of government similar to that of Germany under the Nazis, then John Doe, a very ordinary sort of guy, might find himself employed as a prison guard in a concentration camp, and very highly recommended for his stern measures against prisoners, his businesslike efficiency, and his sincere desire to please his superiors. Right now, in the real world, Doe lives an unnoteworthy life. He works as a security guard at an office building, earns a moderately good salary, is faithful to his wife and kind to his children. But if the changes in his external circumstances are like those that occurred in the thirties in Germany, then without any further change in the dispositions that collectively constitute his moral character, he might routinely perform acts of great cruelty. But in the real world, as opposed to the hypothetical world so easily created by philosophers, he does not appear to be a cruel man. His circumstances give him no reason for it.

Richard Roe shares some of John Doe's dispositions of character, but cruel behavior is often elicited from him even in the actual circumstances in which we all live now. He is, in one rather unfamiliar sense, more "cruel" than John Doe simply because his immoral behavior is more readily produced by his actual circumstances. The more remote and unlikely the circumstances that can trigger acts of cruelty from the person who is predisposed to be cruel, the less cruel, in this sense, that person is. Similarly, my disposition to be a wife-murderer would call for circumstances so extraordinary and unlikely that I can hardly imagine, without the help of a novelist, what they might be. For that reason it is clear that I am much less disposed to be a wife-murderer than, say, Bluebeard or Henry the Eighth.

Of course, the larger number of actual circumstances in which their uxoricidal dispositions would be activated is not the only respect in which these murderous scoundrels differ morally from me, to my advantage. More important, perhaps, is the fact that Bluebeard has dispositions to kill people in circumstances in which I would not kill, but that there are no circumstances, or many fewer of them, in which I would kill but he would not. This clear respect in which Bluebeard and King Henry are *more* savage, ruthless, and cruel than I is not just their bad luck or my good luck in stumbling into, or away from, activating circumstances. Rather that difference, I hope it is clear, reflects a comparison of our separate dispositions to be cruel, namely that their dispositions surface on occasions when mine does not, and are therefore, in that sense, stronger than mine. The important difference between me and Bluebeard (as Walter Sinnott-Armstrong has pointed out to me in his helpful commentary) is not just that Bluebeard "faced actual circumstances which would also have triggered my dispositions" but rather that "he committed murder in circumstances in which [I] would not have done so." Thus my disposition to murder, appraised in this somewhat more familiar way, is less *inclusive*, and therefore not of the same "strength," by that test, as his.

Situational (or "circumstantial") tests of dispositional strength are those that hold the intrinsic characteristics of a persons' character constant and vary the circumstances in which that person might find himself. The more circumstances (total circumstances in some tests, probable ones only in other tests) trigger the disposition, the stronger the disposition must be. The inclusiveness test discussed previously is a way of applying situational tests to the direct comparison of two or more individuals in respect to the strengths of dispositions they may have in common. Now we come to a second type of criterion, the kind that employs, what we can call "intrinsic indicator tests." These criteria supplement the situational tests by holding the circumstances in which a person finds himself constant, and alternating in various ways the character traits themselves. They measure strength by considering how difficult it would be right here and now to "trigger" the disposition, how resistant to inducement it is.[10] The metaphor of the "trigger pull" of a gun, measured in pounds, is useful here if not taken too seriously. According to our first ("situational") test of dispositional strength, discussed earlier, the more remote and improbable the circumstances in which one is disposed to commit a given crime, the weaker that disposition is, and in most cases a weak disposition to commit a given kind of crime can be treated in the same way as a strong disposition *not* to commit that crime. The second kind of measure of dispositional strength (the "intrinsic" test) determines how easily manipulable the disposition is, and may yield the result that although a person has a strong disposition not to commit a given crime (by the situational criteria), that disposition is rather easily overcome by manipulative inducements, in which case it may continue to be appraised as strong by the first criterion, but weak by the second.

It may be less misleading to suggest that there are two senses of the word "strength" involved here, rather than two criteria of strength in the same sense of the word "strength." In any case, the greater the amount of pressure required to move a person's "trigger" to the firing point, the stronger must have been her disposition *not* to commit the crime, and of course, the stronger the disposition not to perform the criminal act, the greater the mitigation of guilt for performing it at another party's initiative. This test, formulated in terms of the triggering metaphor, may be the same measure of dispositional strength and weakness that Aristotle expressed in botanical (not mechanical) metaphors. The more "deeply rooted" and resistant to weakening a given disposition is, the stronger it is. So if my disposition *not* to perform acts of the kind you want me to perform is deeply rooted (perhaps through continual "cultivation" over a long period of time), then my affirmative disposition to perform it is weak, and the mitigation of my guilt considerable. Finally, we can measure the *comparative strength* of a disposition, by determining empirically whether it determines my behavior in more types of situation than other people's dispositions to the same conduct do. This is the "inclusiveness test" discussed earlier. In conclusion, a disposition that is *bad but weak* by any of these quantitative measures, is less guilty than a disposition that is *bad and strong.*

It is useful to have the distinction between intrinsic variations in the strength of dispositions and situational variations, because it can happen, and in unusual cases does happen, that a person acts totally out of character in doing some bizarre or wicked deed, when in fact the dispositions that constitute his character have *not* vanished or changed, but remain precisely the same as they have always been, while the unusual circumstances that can lead him voluntarily to act in so uncharacteristic a way, have come into his situation for the first time in his life.

Moral Luck

Sting victims like Representative Kelly are unlucky that they have become properly subject to adverse moral judgment. Such "victims" of police entrapment have been revealed as seriously defective morally, even though the circumstances in which they acted immorally are so rare that they would almost certainly not have occurred without being elaborately contrived by police who meant to discredit them. In fact the circumstances that led to their immoral conduct were so unlikely to occur that their dispositions to behave as they did in those circumstances might have been unknown even to them, coming as a surprise and a disappointment.[11] Philosophers in recent years, stimulated by the exchange on the subject between Bernard Williams and Thomas Nagel,[12] have rediscovered what they call "moral luck," but that is not the kind of luck, I think, that characterizes the Kelly episode. Therefore, I shall reopen the matter here, and taking the ABSCAM traps as my example, and Nagel's essay as my foil, attempt to explain in what way the desert of adverse moral judgment can, and in what way it cannot, be a matter of luck.

Nagel's basic mistake, it seems to me, is to begin with an unrepresentative set of examples. He seems to focus his attention on a relatively narrow and somewhat unrepresentative class of moral judgments, namely, ascriptions of blame *for* some unfortunate happening or result. These judgments are analogous to, and probably derived from, judgments of civil and criminal liability in courts of law. But even in the noninstitutional settings of ordinary life we have many occasions to make judgments of the form: *B*'s injury was *A*'s fault, or (equivalently) *A* is to blame for the harm done to *B*, or (still equivalently) *A* is responsible (properly subject to adverse criticism, pressure to make compensation, or even punishment) for *B*'s losses. Responsibility judgments of this type, when made informally outside of a legally structured context like a courtroom, can be labeled "moral judgments," but they hardly exhaust the whole miscellaneous class of judgments we call "moral." It is equally a moral judgment to say that *A* was morally at fault in his conduct but fortunately no one was harmed by it. There may be no harm to be compensated, and no clearly evidenced behavior to be punished for, yet *A* is properly subject (liable) to adverse criticism for what he did. A traditional moralist might even say of some other person that he is "morally guilty" or blameworthy for entertaining (say) lascivious thoughts, even though they are not reflected in his conduct, much less implicated in the causation of some specific harm.

Nagel points out that what we are properly held responsible *for* is often in large part a matter of luck. Standard legal examples clearly bear out that claim. A driver who speeds down a residential street at sixty miles per hour will be found guilty of reckless driving and sentenced (perhaps) to forgo his driver's license for one year. But if another driver, differing in no relevant respect from the first, inadvertently runs over a drunken pedestrian who steps carelessly into her path, the driver may be punished for manslaughter and in some places imprisoned for several years. Similarly, we can judge two hypothetical tales of bank robbery in the same way. In one version of the story, the robber aims his gun at a guard and squeezes the trigger with intent to kill or seriously injure him, and he is successful in his undertaking – the guard dies. In the other version, the robber has exactly the same conscious objective, the same intention, and the same motive, and therefore the same degree of "moral guilt,"[13] but the guard moves at the last moment so that the robber misses his target. The first armed robber can be found guilty of first degree murder and in some places executed, whereas the second robber can be convicted only of attempted murder, which is punished much less severely. Nagel is right in pointing out that the difference in severity of punishment in the two cases is entirely a matter of luck, good luck for the first reckless driver and the second bank robber, bad luck for the second reckless driver and the first bank robber.

The reason why luck has so large a role in the law is clear. In cases like these the pronouncements made by courts are judgments of *responsibility for some result*, and the actions of the defendants make only partial causal contributions to that unhappy result. Crucial contributions are also made by other causal factors for which the defendant can take neither blame nor credit because they were beyond his control, such as whether or not the targeted bank guard moves at the last moment, or the drunken pedestrian lurches into the street.

Clearly criminal liability, like civil liability in torts, is partly a matter of luck. But this doesn't show that all moral judgments are matters of luck, which in some places seems to be Nagel's view. Only "blame-for" judgments assigning responsibility for harmful results have been shown thus far to fit the luck model. I don't see how it is a matter of luck whether certain other types of moral judgment correctly apply to one's conduct. It is bad luck that a pedestrian happened to appear in the path of the reckless driver, but it is not bad luck that she was at fault in driving sixty miles per hour. That is, it is not a matter of luck that the latter moral judgment is true of her, at least for any reason we have thus far considered.

There are roughly four ways, according to Nagel, in which "the natural objects of moral assessment" are disturbingly subject to luck, but only two of them are relevant to the theme of this paper, namely those he calls "luck in one's circumstances – the kinds of problems and situations one faces" and "luck in the way one's actions and projects turn out." The other two deal with "luck in how we are determined by antecedent circumstances," and derive from the problem of determinism and free will. We would be wise to set them aside here for they could only lead us down a dark and digressive path.

First, luck in the way things turn out. Will the motorist be charged with reckless driving merely or with manslaughter? Will the bank robber with the murderous intentions be charged with murder or with attempted murder only? Imagine that a "little bit of negligence," through bad luck, leads to a child's death. In a luckier case the same degree of negligence does not lead to anyone's death or injury. Is the luckier negligent party to be judged morally better than his unlucky counterpart? I had assumed that the moral judgments we make in such cases, as opposed to judgments of civil and criminal liability, would restrict themselves to what was within the actor's control, and thus eliminate luck from the moral realm. But Nagel at this point surprises us by making the actor's moral status a function of the whole immediate outcome of his acts, including parts of that outcome that were quite beyond his control. "How things turn out," he says, "determine what [the actor] has done"[14]–whether for example he has endangered, threatened, or harmed another, where the verbs "threaten," "endanger" and "harm" are each the name of a different action. "Actual results," he says, "influence culpability or esteem in a large class of . . . cases ranging from negligence through political choice."[15]

I think that what Nagel says here is true of some legal judgments, but it seems highly paradoxical when said of moral ones. Perhaps the restriction of *moral* praise- or blameworthiness to what is under the control of the actor is part of the *definition* of the moral, or part of the point of the distinction between legal and moral judgments in the first place. Nagel's illustrative examples may seem to him to strengthen his position, but to me they seem closer to a *reductio*:

> If one negligently leaves the bath running with the baby in it, one will realize, as one bounds up the stairs toward the bathroom, that if the baby has drowned, one has done something awful, whereas if it has not, one has merely been careless. . . .[16]

Luck does play a vital role in this example. It is, from the actor's standpoint, a matter of luck whether the baby lives or dies. But it doesn't follow that it is a matter of luck whether the actor is properly judged blameworthy or praiseworthy in his actions, right or wrong, good or bad, conscientious or indifferent. In both the fortunate and the tragic versions of Nagel's story, the actor's conduct is exactly the same. What is different is the unforeseeable contributions of outside factors to the outcome. It can hardly be true of the two actors in these parallel cases that their beliefs, desires, motives, and intentions were identically the same, but that the moral quality of their actions is radically different, one the work of an "awful" killer, the other the product of mere carelessness. Where Nagel is misled, I think, is in his assumption that the *actions* in the two stories are themselves different, contrary to what is actually given in the telling of the story. To be sure, "killing a baby" is, in a sense, a different "action" from merely "endangering a baby," but the actual conduct in the two stories is precisely the same. "Killing a baby" is causing a baby to die through one's faulty behavior; what distinguishes it from merely endangering a baby is the outcome (whether or not the baby dies) and that is at least partially beyond the control of the actor, and therefore a matter of luck. But it is not a matter of luck whether the actor's conduct is truly judged blameworthy.

Even the element of luck that is assuredly enshrined in the criminal law might in principle be subject to removal, or at least modification, without hindering the pursuit of any reasonable aim of that branch of the law. That could be done by eliminating the causal requirement from the definition of all crimes, for example by eliminating the requirement in the law of homicide that the actor's victim *die* as a consequence of the actor's attempt to kill him. It we took that step, then we would never have to decide whether the defendant was guilty in the sense of responsible *for* the harm or to blame *for* the death, but rather only if he was guilty *in* acting as he did, or guilty *of* breaking the law. Perhaps the explanation of the persistence of the causal requirement in criminal law is that there is still a confusion between the aims of criminal and tort law. A terrible harm occurs to someone, and a cry goes up – "Someone must pay for this!" Then a search is undertaken to locate a wrongdoer whose criminal act *cum* criminal mind was "the cause" of the harm, and that party must "pay" for it. In tort law, there typically has to be a causal connection between a given harm and a faulty act that was its cause. The whole point of tort law is to *compensate* innocent parties for the harm they suffered as a consequence of the defective actions of others. If there is no harm to begin with, there is nothing more for tort law to do, since there is no loss to make compensation for.

There may still be something, however, for the criminal law to do, for its function is to punish the parties who perform criminal acts, whether or not harmful consequences actually ensue. To be sure, legislators should enact prohibitory statutes only to prevent types of conduct that tend to cause harm to others. But it does not follow that such statutes should be enforced only when their violation actually causes harm. On the whole, the rationale of the criminal law is that it reduces the net amount of harm all around by discouraging conduct that is in fact dangerous. But it does that job more efficiently when it punishes disobedience and neglects its confused mission of making people "pay" an additional surcharge for actual harms. An alternative worth considering would be to say that tortfeasors are civilly responsible for the harmful *consequences* of their acts and omissions, whereas criminals, as such, are criminally responsible not "for subsequent states of affairs," but for their own criminal actions in disobedience to law.[17]

Another category of moral luck, according to Nagel, is "luck in one's circumstances." "It may be true of someone that in a dangerous situation he would behave in a cowardly or heroic fashion, but if the occasion never arises, he will never have the chance to distinguish or disgrace himself in this way, and his moral record will be different. . . ."[18] World War II ended before my infantry division was ever assigned to combat. I shall never know, therefore, how I would have behaved under fearsome combat conditions. If we assume, for the sake of understanding the philosophical point merely, that my unactualized disposition was to behave heroically under such conditions, then it would have become part of my moral record that in one important arena, at least, I behaved heroically. On the contrary assumption, that I was disposed to act cowardly, or hysterically, or freeze in terror in such activating circumstances, then the judgment that I responded uncourageously to the call of duty would have become a permanent and irrevocable part of my moral record.

But in fact no reference to combat heroism or combat cowardice appears in my moral record because there was never an opportunity to put the matter to the test. It does not follow, of course, that I had no disposition at all in respect to combat duties, one way or the other. I think that probably the correct account of the matter is that I did (indeed that I still do) have some kind of disposition to act in some way or other in extremely dangerous circumstances, but that since those circumstances are now unlikely ever to be actualized, neither I nor anyone else can know just exactly what that disposition is. It may be, as Michael Dummett says,[19] that there simply is no disposition to be known since the only circumstances in which such a disposition could be formed have never become actual. It seems much more plausible to say, however, that there is a disposition to be known if only its activating circumstances would come into existence. However I would have behaved in the invasion of Japan that I was scheduled to be a part of in November 1945, I would have revealed a disposition to respond in *some way or other* to combat danger, whether I charged an enemy machine gun, ran away, or lay down to sleep.

Nagel concludes that in respect to the circumstances that can activate my dispositions, "Here too one is morally at the mercy of fate. . . ."[20] By this he means, presumably, that what appears on one's "moral record" is subject to the occurrence or nonoccurrence of the circumstances that would activate one's disposition, themselves unknown and beyond one's control. What follows from this, of course, is that what ends up appearing on one's moral record is often a matter of luck, not because the existence of the pertinent disposition is a matter of luck, but rather because the occurrence or nonoccurrence of the activating circumstances are. One's "moral record," in Nagel's sense, is part of the historical record. It lists the events, the actions, and omissions, the impelling thoughts and desires that happened in public time and space, not the mere potentialities that might have been actualized in different circumstances, but in fact were not. So when Nagel says that the moral record is partly a matter of luck, he doesn't mean that one's moral character, itself consisting of one's dispositions to act or feel in certain ways in certain situations, is a matter of luck (at least not in virtue of the argument from lucky circumstances that we are presently considering).

We can concede then that it is luck which situations we find ourselves in, and even how our dispositions mesh with unusual or unprecedented circumstances to yield our concrete historical actions. But God, presumably, would see our character dispositions and make accurate moral judgments of them even when they are opaque to all finite vision, and even to the person whose dispositions they are. In general, our actual moral character need have nothing to do with luck. Except for its severe oversimplification, a physical analogy might illustrate the point. In the period before the properties of sugar were well known, a given sugar cube may have been preserved in its solid state for many decades. Even if that sugar cube were never placed in water, it would surely have remained soluble, even though the liquid circumstances that foster its dissolving were not present, and none of the people who *were* present had any reason to suspect how it would behave in water.

Nagel says that "we judge people for what they actually do or fail to do, not just for what they would have done if circumstances had been different."[21] We do praise and blame people's actual behavior far more often than we assess their character as a whole, but that should not be surprising given how many serious practical issues ride on our judgments of actual behavior. Moreover, the relative infrequency of character assessment is partly to be explained by our humility in the face of what is highly speculative and uncertain. It is only because we lack confidence in our knowledge of a stranger's disposition that we surmise that in the country churchyard of Thomas Gray's poem lies "some Cromwell, guiltless of his country's blood."[22] Very likely there were contemporaries of Cromwell whose lives (moral records) were far better than his, and who were innocent of murderous atrocities like those of Cromwell's troops only because they lacked the opportunities, in the circumstances that prevailed, to do the same. We cannot know of any given person that such a description fits him, lacking as we do evidence of such potential in his behavior, but it is an intellectually titillating speculation that there were such disguised moral monsters, as there may have been in our time unsuspected potential Hitlers (though surely not many). Indeed given the "banality of evil"[23] that Hannah Arendt gleaned from the Adolf Eichmann story, there may be thousands of moral Eichmanns in towns and villages all over the world – luckier people than the German organizer of death camps, but in the eyes of God of no better disposition.

Our modest conclusion thus far is that there is such a thing as "circumstantial luck," but little reason to posit a corresponding "dispositional luck."

That is to say that it may sometimes be a matter of luck what circumstances we find ourselves in (what background conditions – physical, biochemical, social, political – as well as what conduct initiated by other persons in a manner requiring our response), but that rarely if ever are our dispositions to act in certain ways in certain circumstances equally based on luck. So what ends up appearing on one's moral record is often a matter of luck, not because the existence of the pertinent disposition is a matter of luck, but rather because the occurrence or nonoccurrence of the activating circumstances is. Moreover, the word "moral" is used misleadingly if it is made part of the label ("moral luck") for the unforseeable variation in circumstances that is a matter of luck, but has nothing to do with morals or with character. It is the dispositions themselves that ground moral judgments about our character, and these are to large degree subject to our own control. The presence or absence of activating circumstances does not in itself determine our moral merit or demerit, and a good thing too because they *are* largely a matter of luck. If I am right in this analysis, the word "moral" goes with our dispositions but not with our circumstances, while the word "luck" goes with our circumstances but not to the same degree with our dispositions. It seems to follow that there is little room amid these concepts for the hybrid idea of "moral luck."[24]

More about Dispositions

Dispositions are ascribed to people by sentences of the form — Usually in circumstances *C,* actor *A* does *x.* Spelling out the circumstances and qualifications, the probabilities and exceptions can make such statements complicated indeed. Some conditional statements of this general form are too complex and specific to be definitions of single-word propensities like the names of virtues (honesty, benevolence, calmness under pressure, etc.), or flaws of character (dishonesty, self-centeredness, excitability). Moreover, we are reluctant to call such a singular disposition a "tendency" or an "inclination," not to say a "habit," because these words all suggest that the activating circumstances mentioned in the antecedent clause of the complex conditional sentence are frequently encountered, whereas some of them are only infrequently or even not at all encountered by the person who may truly be said to have the disposition in question. It may be true of such a person that whenever she is in complex circumstances *C,* or if ever she were in such circumstances, she does or would do *A,* even though she has never ever been in *C* and is never likely to be in *C.* In that event we might have no way of learning whether the person really does have the disposition in question or even the general inclination associated with it, except perhaps by making very shaky inferences from the dispositions we do know she has, and even that may give us very little guidance. We can say that God, being omniscient, does in fact know whether the statement attributing the disposition is true even though the activating circumstances mentioned in the antecedent clause have never existed, and even though they may never exist in the future. In other words, there is a fact of the matter here even if it cannot be known by mere mortals. "If ever she were in *C*, Jane would do *A*" may be true or false though we may never know which.

A person can be said to suffer bad luck when his disposition to do something morally wrong is activated by circumstances that are extremely unusual, and which before they occurred were correctly judged extremely remote and unlikely. It may be true of me, for example, that when and only when huge volcanic eruptions fill the skies with darkening debris for at least twelve months do I become inclined to commit mayhem, counterfeiting, and embracery,[25] and that I will in fact commit that miscellany of crimes at the point when the skies lighten again. This is such a bizarre disposition that there would be no discovering it by inference from my "analogous" dispositions, so probably it could be known only to God. It certainly could not have been known, before its activation, to *me*. Now let us suppose that the skies darken to the requisite extent and for the requisite period and then begin to lighten again. Without showing any (other?) signs of madness I commit mayhem on my neighbor; I make crude copies of $100 bills and distribute them as widely as I can; then I seek out a criminal trial, befriend a juror, and attempt wrongfully to influence him. I am, of course, captured, convicted, and sentenced to severe punishment. I was unlucky in that conditions that occur only once every four or five centuries, on the average, occurred in my time, and led me to commit acts of undoubted wrongfulness.

I am not "unlucky," however, in having the wrongful disposition in the first place. That would be an instance of what Nagel and Williams called (bad) "moral luck." It would indeed be bad moral luck, for instance, to have a moral record containing instances of behavior deemed immoral because of factors beyond one's control and contrary to one's intention. Similarly it would be bad moral luck, so called, to have a disposition properly deemed blameworthy even in those of its aspects that are beyond one's control and contrary to one's predominant values.

But, putting aside all of that, what *is* unlucky is that one should stumble into the unlucky circumstances, rare as they may be, which unknown to the person in question, can actualize one of her wrongful potentialities, which in turn may also have been unknown to her. It is important to note, however, that the disposition remains a wrongful one, quite apart from the bad luck suffered by its possessor in its unlikely activation. A permanent bachelor's disposition to be unfaithful to his wife, if only he had one, is an immoral character trait even though it never can lead to an immoral action of the kind we call "adultery." What makes that disposition immoral is something quite other than that which makes it difficult or impossible to exercise. A similar judgment applies to a permanent civilian's tendency to be a cowardly soldier in combat. Both the civilian's cowardice and the bachelor's infidelity, however, may be particular instantiations of general failings of an analogous kind, so that the bachelor might be more inclined than most to betray people who trust him, and the civilian might be inclined to cowardice in a dozen other contexts, such cowardice leading those who know of it to "expect" that if he were a soldier he would be cowardly in that context too. The reason why the earlier example of the linkage between volcanic explosions and the counterfeiting of currency was so bizarre and incredible is that we cannot infer from it any more general kind of failing of which this strange one is only an instance.

Some Moral Judgments

One argument meant to minimize the guilt of a person who has been successfully manipulated by police agents into voluntary behavior that would seem "out of character" for her begins with a distinction between the *initiative* taken in an interpersonal transaction, and the degree to which that action reflects the character of the actor. Typically, the strategy of an entrapment attempt as well as the decision to undertake it begins with another party. The behavior that eventually seals my doom, if I am the one induced into action, originates in the mind of the police agent. But for her initiative it would probably never have occurred to me at all to do the acts she was trying to get me to do. On my own, I am completely innocent. Of course, she didn't use force or the kind of deception that would vitiate the voluntariness of my response to her. But she did bring into existence the condition mentioned in the antecedent clause of the conditional statement that defined my disposition, and that took a great deal of doing. The action she desired from me is linked to very few dispositions that anyone can know of in advance, and in this case the triggering condition is very unusual, of a kind infrequently and improbably encountered by me. Since it was her initiative and not my disposition that is the critical element in the explanation of my criminal action, that action reflects more on her than on me. My doing, we might say, was primarily *her* doing, since the initiative in the first instance was entirely hers.

That argument, though on to something, does not quite succeed. It is not an automatic reducer of the guilt in someone's action to show that the original initiative for that action was taken by someone else who undertook to persuade the acting party by noncoercive manipulation. Indeed, the greater part of our morally interesting conduct, and our challenging moral dilemmas, requires us to decide how to respond to the initiative of others. You offer to sell; I decide whether to buy. I propose that we get married; it is now up to you to consent or not. You offer me a financial inducement at just the time when I need the money most; I wonder whether to accept your terms or not.

You ask for my signature on a political manifesto; I must decide whether my agreements with it are more or less important than my disagreements. In none of these commonplace examples do I later have the "excuse" that in the beginning it was another person's idea, not mine, and that even though I responded as I did to her initiative of my own free will, the responsibility or guilt for what I did must be hers, since she started the whole thing. In short, locating the initiative in the other party does not automatically, with nothing more, provide the voluntary actor with an exculpatory excuse.

The point to emphasize is that while the other's initiative is not the whole story, it can be a crucial part of the story. Showing that your initiative was behind my voluntary act does not at all by itself reduce my guilt. What must be shown is that the disposition of mine that your initiative was meant to activate had certain characteristics. If I were known to be a skilled and experienced bank robber, you might propose that we do a job together, and split the profits. Moreover, you may have a detailed proposal for my consideration in which you identify the bank, propose the date and time of day, provide the tools, and so on. If I agree and then get entrapped, because unknown to me, you are now a paid *agent provocateur*, I will certainly point out that your initiative was behind the criminal act, not originally mine. But that true observation will do me little good unless I can show that the predisposition to such acts, which I obviously had, was a very weak one. Indeed, the weaker the disposition, one might say, the less dangerous is the criminal, and the less guilty is any criminal act that another party induces him to perform. That rough rule of thumb is the best statement we can come up with of the mitigating effect of another party's initiative.

We come finally to the problem of private entrapment. When a sexy person entices another person into the commission of a crime, the criminal actor cannot plead, either in a criminal court or in "the court of heaven," that the enticement excuses his otherwise criminal behavior *unless* – the sexy person is a plain clothes policeman or policewoman. Why should there be this exception? The entrapped victim, it would seem, is equally guilty, since the only difference between the two cases is one unknown to him (or her). Two answers are often given. The first is the answer of "the objective theory of entrapment." According to this view, the person entrapped by police agents should not be convicted at all (in the most flagrant cases not even tried) not because such a person is without moral blameworthiness or guilt, but simply because the police in a democracy must be kept honest and trustworthy. This rationale is essentially similar to that frequently proposed for the exclusionary rule which renders inadmissible, evidence seized in violation of the fourth amendment's stricture against unreasonable searches and seizures.

The other answer contains some elements in common with the first, but it is somewhat more interesting to the moral philosopher. When the state has license to test the virtue of each and any citizen, randomly selected, then for a substantial number of such citizens the circumstances that can trigger the criminal predisposition are artificially created, and would not otherwise ever have arisen. This is quite another thing than randomly setting up police roadblocks to test drivers for drunkenness. People who are disposed to drive after immoderate drinking are punished not just for that disposition but for actually driving while drunk. So far as I know, no plain clothes police agent has ever been hired to offer them free drinks in their parking lots as they climb into their cars for the trip home from work, thus tipping the odds against the drivers they successfully persuade to accept the gift. That is the nub of the complaint against entrapment practices when they are objectionable. People who might otherwise never be in trouble with the law are convicted and punished anyway because and only because the police have intervened, without sufficient cause, in their lives. They have set out deliberately to alter the odds that are otherwise so unthreatening to law-abiding citizens.

If police have the power *randomly* to intervene in people's lives just to determine whether they have wrongful predispositions, then we are all subject to a greater danger from those who are supposed to be our protectors. If they proceed in a genuinely random way, as their colleagues, for example, in the case of the drunkenness-testing roadblocks, then they can exercise their power to intervene in our lives without any evidence, even without any suspicions or hunches of wrongful doing on our part. That would appear to be a violation of our liberty, according to our traditional interpretation of the constitutional limits on police activity. But the increases in police power can be even worse if they are *not* random, as when, for example, they are put to political uses through the virtue-testing of the leaders of dissident political factions. In that case the police powers are being used to harass, and when a given dissident is not merely harassed but actually entrapped, she is treated even worse.

It may be more palatable, however, to put responsible government officials, federal judges, and congressmen, for example, to the test of honesty, since disposition-triggering circumstances (temptations) occur more frequently in their offices, and the chance of being put to the test of honesty is already high. Moreover, occupants of these positions have special responsibilities to be trustworthy, since they are the makers and appliers of law, and securing assurances of their reliability, even in the absence of any suspicion of their unreliability, is more pointedly in the public interest. It is plausible to tell them that being subject to a special scrutiny is a reasonable price to pay for their power. That argument is weaker when applied to a less powerful group (like college professors) and downright pernicious when directed only at political dissidents, religious heretics, and so on.

Most of these reasons against permitting police to test for virtue, and thus alter the odds that testees would ever become prisoners, apply more or less to the private seductress who, without using deception, force, or violence, lures another party, who might otherwise never have committed a crime, into being her partner, not for the purpose of betraying him to the police, but rather for her own private criminal purposes. So it seems that the ultimate difference between private seduction to crime and police enticement to crime is, after all, found mainly in our need to control police power in a democratic state. But this solution cannot be the whole of it. It really is more satisfactory knowing that the "natural odds" against my ever committing a crime and going to prison for it (now extremely high) – odds that include among their data the chances (now quite low) that I will be lured into a crime by a seductress for her own purposes – remain unaltered by the manipulations of the very state agency that is charged with my protection. I, for one, will be happy to take my chances with the natural odds, though unfortunately, in this imperfectly reliable world, even they give me no guarantee.

NOTES

1. See my *Harm to Self* (New York: Oxford University Press, 1986), pp. 269–316.
2. *Sorrells v. United States* 287 U.S. 435 (1932).
3. Justice Roberts, as quoted in W. R. LaFave and A. W. Scott, *Criminal Law*, 2nd ed. (St. Paul: West Publishing, 1986), p. 421.
4. Leo Katz, *Bad Acts and Guilty Minds* (Chicago: University of Chicago Press, 1987), p. 155.
5. For a discussion of this metaphor, see "Causing Voluntary Actions," in my book *Doing and Deserving* (Princeton, NJ: Princeton University Press, 1970), pp. 152–86.
6. *United States v. Kelly* 748 F. 2d 691.
7. Katz, *Bad Acts and Guilty Minds,* pp. 157–8.
8. Ibid., p. 158.
9. Gerald Dworkin, "The Serpent Beguiled Me and I Did Eat: Entrapment and the Creation of Crime," *Law and Philosophy* 4 (1985): 33.
10. We can be speaking now of strong dispositions *not* to commit a crime or, equivalently, of weak dispositions to commit the crime.
11. Discovering our formed dispositions may frequently come as a surprise to us, at any time of life. Self-knowledge may require self-discovery even of those aspects of our character that we have ourselves created by our voluntary actions and omissions. Cf. the following sentence in Lewisohn's novel, p. 78: "Then a strange thing happened. Herbert, who didn't know he was irascible, felt a wave of violent irritation rise up in him."
12. Bernard Williams, "Moral Luck," *Proceedings of the Aristotelian Society*, suppl. vol. 50 (1976): 115–35, and Thomas Nagel, "Moral Luck" in his *Mortal Questions* (Cambridge: Cambridge University Press, 1979), pp. 24–38.
13. Justices Rehnquist and Scalia apparently regard such stories of bad luck in the criminal law, not as "moral paradoxes," but as the plain deliverances of common sense. See Scalia's opinions in *Booth v. Maryland,* 482 U.S. 496 (1987), and *South Carolina v. Gathers,* 490 U.S. 805 (1989), and Rehnquist's opinion in *Payne v. Tennessee* 11 S.Ct. 2597 (1991). In *Payne,* Justice Rehnquist quotes the example (from Scalia) of the two identical bank robbers and adds, with hearty approval, that "under the facts of *Tison v. Arizona,* a bank robber who acts recklessly and kills someone can be put to death while a bank robber who acts recklessly and kills no one cannot." Quoted from "When the Law Bows to

Politics: Explaining *Payne v. Tennessee*'' by David Dow in *U.C. Davis Law Review* 26 (1992): 165.

14. Thomas Nagel, ''Moral Luck,'' in *Mortal Questions* (Cambridge: Cambridge University Press, 1974), pp. 29–30.
15. Ibid., p. 30.
16. Ibid., pp. 30–1.
17. The preposition ''for'' seems to have a different sense in the judgment that a person is responsible for the *harm* he has caused than it bears in the charge that a person is responsible for his wrongful *actions*. We can detach actions from their results so that it makes sense to say that one is the cause of the other, but we cannot detach the actor from his actions, so that he can be labeled the cause of those actions. We don't cause our actions; we just do them.
18. Nagel, ''Moral Luck,'' pp. 33–4.
19. Michael Dummett, *Truth and Other Enigmas* (Cambridge, Mass.: Harvard University Press, 1978), pp. 14–16.
20. Nagel, ''Moral Luck,'' p. 34.
21. Ibid.
22. Thomas Gray, ''Elegy Written in a Country Churchyard,'' as quoted by Nagel, ''Moral Luck,'' p. 34n. The lines in question are:
    Some Mute Inglorious Milton here may rest.
    Some Cromwell, guiltless of his country's blood.
23. Hannah Arendt, *Eichmann in Jerusalem: A Report on the Banality of Evil* (New York: Viking Press, 1963).
24. I do not mean to deny that there might be an element of luck in the development of one's moral dispositions, as also of one's abilities, capacities, and the like. The usual argument for ''moral luck'' in this dispositional sense, however, is that our dispositions come to us already determined, for better or worse, and are not the products of our own efforts and deserts. That is a complex argument that I have prudently avoided considering here, concentrating instead on Nagel's account of two other types of ''moral luck.'' (See Nagel, ''Moral Luck,'' pp. 29–30.) There is enough stimulation in Nagel's other categories fully to occupy a paper of this length. To take a further stand on the question of antecedent determination of dispositions would involve us in the complex of puzzles about free will, the analysis of the concept of luck, Rawls's natural lottery theory, and Aristotle's and Sartre's divergent accounts of how it is that we can ''choose our own characters.''
25. This word refers to the old common law crime of wrongly influencing juries in certain specific ways.

11

Marcus and the Problem of Nested Deontic Modalities

NUEL BELNAP AND PAUL BARTHA

1. The Parking Problem

Marcus's short paper "Iterated Deontic Modalities" (1966) is re-readable, and recompelling. She argues that standard deontic logic makes too little sense of putting one deontic modality inside another. Even now deontic logicians shy away from nesting their modalities one within the scope of another. One of the reasons apparent from her argument is this: *Direct* iteration is of little use. If, however, one takes the trouble to insert agentive constructions after the deontic operators, then embedding suddenly becomes a powerful analytical tool.

Perloff and one of us (Belnap) have refined this suggestion by taking the agentive construct to be "modal," that is, to map sentences into sentences. We think that the modal move is critical. The plan here is to explore this idea by intertwining certain themes from Marcus's work on "general-purpose" modal logic with themes from the modal logic of agency in its deontic application. In fact the nesting of deontic modalities is taken up only quite late in this study. Before then we shall be considering – all too briefly – the representation of obligations and prohibitions, moral dilemmas, the serious use of quantifiers in modal logic, Barcan formulas, and problems of extensionality or its absence.[1]

As a device for the remainder of this paper, we shall concentrate on an unpretentious example from Marcus 1966 (p. 580). There is much to be said for unpretentious examples. After all, "7 + 5 = 12" was good enough for Kant. The Marcus example is, however, more serious than mere arithmetic:

(1) Parking on highways ought to be forbidden.

With this example Marcus is challenging all deontic logicians. Deontic logic, she says, could be a "misleading technical exercise" (p. 582) if it treats (1) as a simple iteration of an impersonal 'ought' and an impersonal 'forbidden'. That is, the following symbols (2) won't do for (1):

(2) $O(F(p))$.

Marcus, with conclusive arguments that we shall not repeat here, pins the blame on a failure to keep separate the *evaluative* modal family headed by 'ought to be' and the *prescriptive* modal family headed by 'ought to do' or 'obligatory'. With reference to the prohibition expressed by 'forbidden', let us leave aside the 'ought to be', turning instead to the 'ought to do'. Marcus endorses this turn. She says

> It seems to me that the least problematic reading of the deontic operators is the one which fits the use of 'obligatory', 'forbidden', 'permitted', as they occur in connection with rule governed conduct. Such a restriction to prescriptive use has the advantage of allowing us to be clearer about the semantical interpretation of deontic statements. (Marcus 1966, p. 581)

She is suggesting that we can best use deontic logic if we focus on the 'ought to do'. In advancing our understanding of the 'ought to do', Marcus, citing K. Baier, suggests the following:

> we may take 'It is obligatory that *A*' (where '*A*' is an appropriate statement of the sort '*x* does *w* at *t*') as meaning that *A* is entailed by the set of rules and standards. (Marcus 1966, p. 582)

This is a rich suggestion. A good deal of the remarks to follow can be construed as employing the modal logic of agency in order to follow it. In doing so we are going to concentrate on (1) to the exclusion of any more gripping examples. If you start to fall asleep, please substitute an illustration filled with dangerous and dreadful actions of the most excitingly electrifying sort. You know, something like 'buttering the toast in the bathroom'.

2. The Form of Obligations

The analysis of (1) proceeds from the inside out. We shall construe (1) as containing the following as a grammatically proper part:

(3) Parking on highways is forbidden.

Most of the work of this paper will consist of articulating the part (3) of (1), using rather simple deontic concepts in combination with the logic of agency. Marcus would hold that even before we reach semantics, the underlying *grammar* of a typical prescription is like this:

(4) It is obligatory that (*x* does *w* at *t*).

This suggestion, itself unpretentious, is worth taking, and has in practice guided our work. In fact the suggestion is so helpful that it is easy to overlook the following: When some philosopher says, '*x* does *w* at *t*,' we have no idea over what the variable '*w*' is supposed to range. In other, more pretentious, words, we don't – even after decades of the most arduous first-order work – have an 'ontology of actions' that is at once useful and rigorous. Nor, of course, does Marcus, the consummate modal logician, suppose that we do.

Enter the modal logic of agency. Perloff and Belnap call their version[2] "*stit* theory," or seeing-to-it-that theory, because it concentrates on the modal construction

α sees to it that Q,

which in most of their work so far has been abbreviated as

[α *stit:* Q].

What *stit* theory contributes is an illuminating framework in which one isn't constantly having to worry about turning actions into *things* to name or to quantify over. *Stit* theory suggests instead that the grammar of a prescription is this:

It is obligatory that α see to it that Q,

or, in more idiomatic English,

(5) α is obligated to see to it that Q.

Three ideas lie behind this formulation, ideas that we should like to mention here without defense, just for background. The first is that the place of Q can meaningfully be filled by any sentence whatsoever. This is the power of exactly that modal point of view championed and defended by Marcus against various forms of "first-orderism."

The second idea is that the place following 'it is obligatory that' can *only* be occupied by an agentive, of which the *stit* sentence is both paradigm and normal form. Perloff and Belnap call this the Restricted Complement Thesis. This thesis gives technical bite to Marcus's distinction between the evaluative and the prescriptive. That is, it forces us to keep the *prescriptive* 'ought-to-do' separate from the *evaluative* modal family.

The third idea is this. In contrast to the way everyone waffles when it comes to an ontology of 'actions', we are in firm possession of a clear and useful semantic account of 'α sees to it that Q'. This account also shares convictions with Marcus 1980. Two are paramount. The first is that obligation and agency are bound up in complex ways with possibility. The second is that you cannot get away from occasions and times. This is witnessed by the 't' in (4), and made explicit in passages such as the following.

'Ought' is indexical in the sense that applications of principles on given occasions project into the future. They concern bringing something about. (Marcus 1980, p. 135)

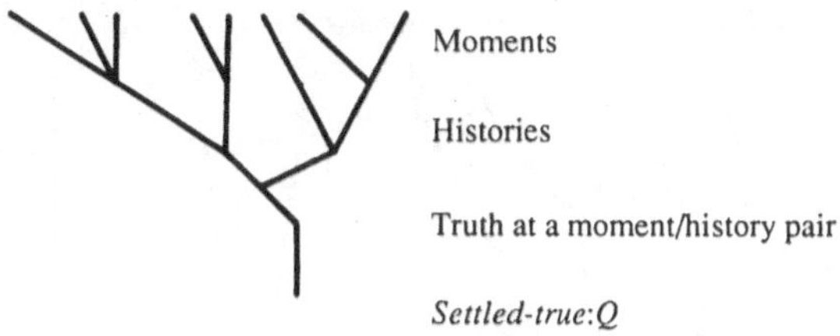

Figure 1. Branching time.

Together these ideas help suggest the development of an independent semantics of *seeing to it that.* The *stit* proposal relies on the Prior–Thomason theory of *branching time*, pictured in Figure 1. The diagram lists the key concepts, for which it is necessary to refer the reader to, for example, Thomason 1984 or to the papers mentioned in note 2. We only cite the semantic account of settled truth.

1. DEFINITION. *Settled-true*: Q is true at a moment/history pair m/h just in case Q is true at m/h_1 for every history h_1 passing through m.

Use of branching time as a foundation presupposes answers to difficult and important questions about actuality and possibility, answers that Marcus roundly challenges in her papers Marcus 1977 and Marcus 1985–6. Even to begin discussing any of these matters, however, would mire us without hope of progress. We therefore call on your charity. Please excuse us on this occasion – but not forever – from defending branching time, which we certainly consider to be plausible, but which others with equal certainty consider to be a doubtful dream.

To the basic picture of our world provided by branching time is added the concept of an *agent,* and, above all, a rudimentary theory of *possible choices.* In this context Perloff and Belnap have considered one expression of agency, *stit,* at length and another, *dstit* – which was suggested independently by Horty and von Kutschera – only in footnotes. When one of us (in Bartha 1991) applied the modal logic of agency to deontic concerns, he chose to idealize with *dstit,* which is in some respects simpler. Here, partly because we are building on Bartha 1991, we also use *dstit.*[3] For the semantics of *dstit* a picture will have to suffice: See Figure 2.

The diagram in Figure 2 is a blow-up of a single moment m, indicating all of the histories that pass through m. The boxes indicate how the histories through m are divided into four possible choices for the agent α. The idea is that α is unable to distinguish between histories that belong to the same possible-choice box. The lettering tells you for each history h through m

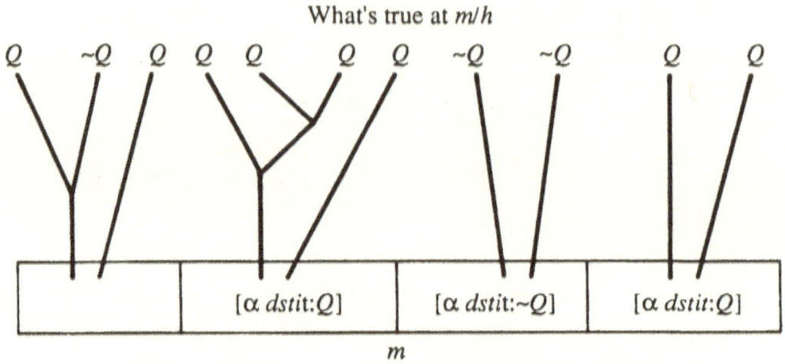

Figure 2. Choices for agent α at moment *m*.

whether or not Q is true at the moment/history pair m/h. The semantics of *dstit* is then this:

2. DEFINITION. [α *dstit*: Q] is true at a moment/history pair m/h if two conditions hold. The *Positive Condition* says that Q is true at every pair m/h_1, where h_1 and h are in the same possible choice for α at m. The *Negative Condition* says that Q is not settled true at m; that is, Q is false at some pair m/h_1 with h_1 passing through m.

The picture indicates, for instance, that [α *dstit:* Q] is true throughout the second and fourth choices for α at m. As you can see, the idea is not complex.

A sentence Q is 'agentive in α' if Q can be paraphrased as [α *dstit:* Q]. For example, let

(6) $P\alpha x \leftrightarrow$ α parks on highway x.[4]

Then $P\alpha x$ is agentive because – or, better, to the extent that – it comes to the same thing as 'α sees to it that $P\alpha x$':

3. POSTULATE. $P\alpha x \leftrightarrow$ [α *dstit:* $P\alpha x$].

We do not propose that the longer sentence is an 'analysis' of the shorter one, but only a useful 'normal form'.

3. The Anderson/*dstit* Simplification

So much for the "bringing it about" part of the Marcus account of prescriptions. What about the normative part? As noted, her thought is to spell this out as entailment by a set of "rules and standards." Bartha 1991 suggests that we take our bearings in this extremely complex area by use of the Andersonian 'sanction', the bad thing, represented by a propositional constant,

S, indexed by agents, with a different 'sanction' S_α for each agent α.[5] Think of S_α as true at just those moment/history pairs at which the particular agent α has not lived up to the rules and standards. Then on Bartha's proposal we can analyze (5) as follows:

4. DEFINITION. *Obligated:* [α *dstit:* *Q*] ↔ *Settled-true:* (~[α *dstit:* *Q*] ⊃ S_α).

That is, at a particular concrete moment, α is obligated to see to it that *Q* just in case in every history through that moment, either α sees to it that *Q* or else the bad thing happens. Observe the following analytically decisive point. If you take Definition 4 as a definition, then it cannot be understood as a definition of '*Obligated:*' in isolation. It only makes sense as a definition of the compound *Obligated:* [α *dstit:* *Q*]. Definition 4 gives no meaning whatsoever to *Obligated:* *Q* for nonagentive *Q*. That this should be so conforms perfectly to the Restricted Complement Thesis. It is precisely what makes this account of obligation prescriptive rather than evaluative.

Note that there is no talk of globally and impersonally ideal worlds or histories. It's just that at that moment you are supposed to do what you are supposed to do, and if you don't, you are in trouble. Bartha 1991 touches on more than one inadequacy of this proposal, but these we pass over instead of pause over. Instead we go on to relate Definition 4 more closely to Marcus motifs.

4. The Form of Prohibitions

A first and interesting problem is the representation of prohibitions. Marcus 1966 notes that " 'forbidden' is more clearly prescriptive of action" (p. 590) than it is evaluative. If the rules forbid you to park on highway *x*, they are prescribing an action. But which action? The trouble is that ~*Pαx*, that is, 'α is not parking on highway *x*' does not describe an action. It describes a state. For instance, α is in that state when sleeping. As Marcus 1966 says in another case (p. 580), "How can it be *forbidden* that such a state of affairs prevail unless it is part of a blueprint for creation?"

It's confusing. 'Forbidden' is supposed to prescribe, and to prescribe is to prescribe action, but since 'not parking' is not an action, it looks as if there is no action to be prescribed. At this point one might think of taking prohibition by a set of rules to be a second primitive notion, to be added to prescription. *Dstit* theory has, however, an absolutely satisfying solution to this puzzle. It goes like this. We know that what is *forbidden* by the rules is that α park on highway *x*. The problem is to say what is *prescribed* in this case, if we can. We already know it won't do to say that it is the state of not parking that is prescribed. *Dstit* theory solves the problem as follows: It says that what is prescribed is that α *see to it that* α does not park on highway *x*. It can do this because although you cannot 'negate an action' (and keep your sanity), you can certainly negate any sentence, including any agentive sentence. And any sentence can be the complement of the *dstit* construction.

More generally, the proposal is this:

5. Definition. *Forbidden:* [α *dstit: Q*] ↔ *Obligated:* [α *dstit:* ~ [α *dstit: Q*]].

That is, α is forbidden (by the rules) to see to it that *Q* if and only if α is obligated (by the rules) to see to it that it is false that α sees to it that *Q*. The key point here is that this obvious and elementary truth cannot be said, or said clearly, without the grammar of *dstit*. To be able to formulate this striking definition of a prohibition as an obligation to refrain, you need to be able to embed a negated *dstit* inside of a *dstit*. It is precisely this capability that seems to be absent from an ontology of 'actions'. This power of embedding, which belongs to the very idea of modality, needs to be exploited. You cannot exploit it if you are ashamed of modalities. That is one of the great themes of Marcus's work: Instead of treating modalities as an embarrassment in the fashion repeatedly preached by Quine, you should use them for what they are worth, which is plenty. Use them by themselves, with quantification, with identity, or with whatever helps.[6]

Combining Definition 5 with Definition 4, we have the following symbolizations and equivalents.

Forbidden: [α *dstit: Pαx*]: α is forbidden to park on highway *x*.

↔

Obligated: [α *dstit:* ~[α *dstit: Pαx*]]: α is obligated to see to it that α does not park on highway *x*.

↔

Settled-true: (~[α *dstit:* ~[α *dstit: Pαx*]] ⊃ S_α): It is settled true that if α does not see to it that α does not park on highway *x*, then S_α.

That's a lot of words. One can better see into the proposal by a combination of cases and pictures. Take a particular concrete moment, *m*, of concern to α and to the rules. There are three major cases, the first two of which are throwaways.

No-Good-Choice Case. Since Bartha 1991, in contrast to Anderson, has put no contraints on the sanction, there is the case in which S_α is settled true at *m*. In the no-good-choice case all is obligatory for α, but also all is forbidden. There is no available choice for α that avoids the sanction. Nothing is permitted, where, as usual,

No-good-choice case

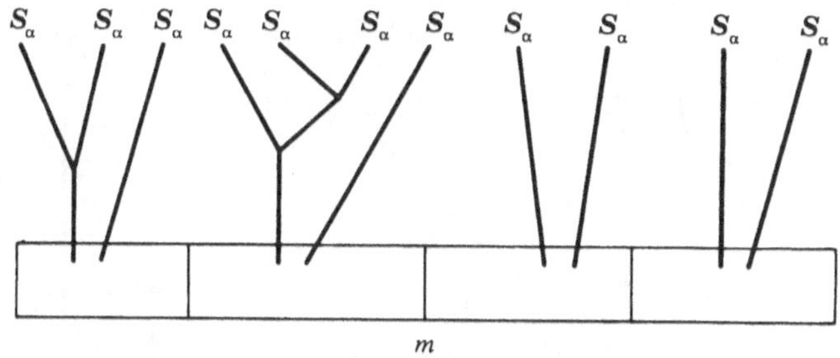

For α, all is forbidden and all is obligatory!

Figure 3. No good choices for α at *m*.

6. DEFINITION. *Permitted:* [α *dstit: Q*] ↔ ~*Forbidden:* [α *dstit: Q*].

The no-good-choice case is pictured in Figure 3. What to make of the different decisions, Anderson's requiring that the sanction not be universal and ours making no such requirement? We think that both decisions are plausible, and that the point of view of Marcus 1980 explains why. Anderson's sanction applies to entire worlds or perhaps histories. It would be strange indeed if every world or every history were deemed unacceptable. In this case there would clearly be nothing left for deontic logic to worry over. As Marcus puts what we take to be the same point, if a set of rules is inconsistent in this strong sense, then it "provides *no* guide to action under any circumstances" (p. 129). This makes reasonable Anderson's postulate that the sanction is not inevitable.

On the other hand, when we mark as 'sanctioned for α' every history through a particular moment, that sad situation is an extremely local affair. It represents only that in that particular circumstance, there is nothing acceptable to do. This is what Marcus urges to be quite another thing. To deny that such a circumstance could obtain is, as she says, to deny something real, namely, that a moral dilemma can arise in a particular unfortunate circumstance. So, we conclude, the two choices are both in accord with Marcus's account, and they are both right.

No-Can-Do-Case. In the other throwaway case, at *m*, α cannot see to it that *Q*: It is settled true that α does not. Nothing α can choose at *m* counts as choosing to see to it that *Q*. No such choice is in fact open to α at *m*. There are two subcases. In one of them the previously discussed no-good-choice case obtains, and there is nothing more to say: All is obligatory and all is forbidden for α, regardless of what the agent can do. In the other subcase, which is new, S_α is false at *m*/*h* for at least one history *h* through *m*. We will call such a case a *proper* no-can-do case. Figure 4 pictures two of them.

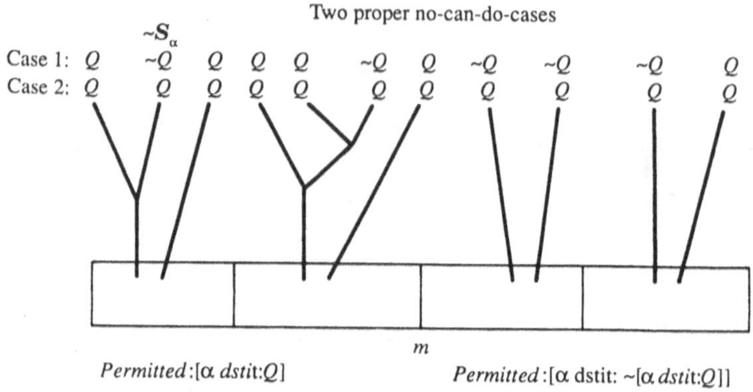

Figure 4. Two proper no-can-do cases.

You can easily calculate that in a proper no-can-do case, Definition 4 neither obligates nor forbids the agent to see to it that *Q*, which sounds right when the agent no can do. Of course it follows that in the proper no-can-do case the agent is *permitted* to see to it that *Q*, and also permitted to refrain from seeing to it that *Q*, which sounds peculiar given that the agent is powerless with respect to *Q*. But it sounds peculiar, we think, only because in ordinary English we confuse permission and power.[7] For example, we often use the same modal verb, 'may', for both. When we reflect, however, we easily see that in fact there are numerous situations in which an agent is permitted or authorized to do what in fact is not in their power to do. The agent α might be permitted to park at any meter in the entire state of Connecticut, even when α is enjoying Oregon. The proper no-can-do case highlights this.

The Normal Case. In the central case of interest, the sanction S_α is false on at least one history through *m*, and there is at least one choice on which α sees to it that *Q*. We will call this *the normal case*, even though it certainly does not happen with much frequency. The normal case is pictured in Figure 5. In the normal case, α can see to it that *Q*. In the normal case, the sanction is not inevitable.

It is a calculation of *dstit* theory that in the normal case, though not at all in general, 'seeing to it that' and 'refraining from seeing to it that' are contradictory.[8] (The sanction is irrelevant to this calculation; one needs only can-do.)

5. Generalized Prohibitions

Let us pass on to the first of the two generalities ingredient in (1). It is not just one highway on which α is forbidden to park. In fact,

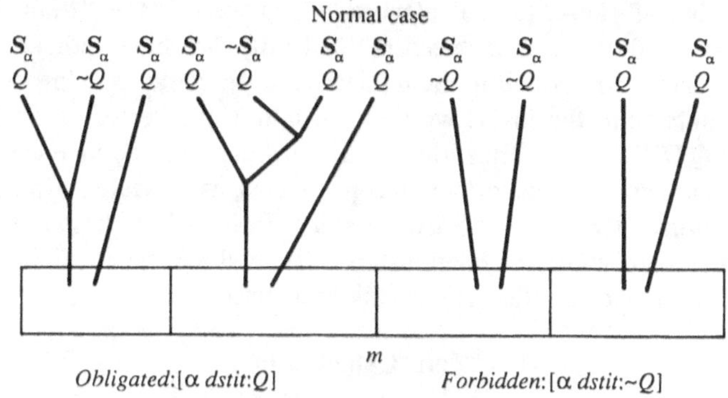

Figure 5. The normal case.

(7) *Each* highway[9] is one on which α is forbidden to park.

In other words, we might say,

(8) α is forbidden to park on *a* highway.

(Or, put more ponderously in order to nail down narrow scope for the quantifier: It is forbidden that α park on *a* highway.) So, obviously, to express that parking is forbidden on highways we need a quantifier. But which quantifier, and where should it go? The universally quantified prohibition (7) and the prohibition of an existential (8) sound sort of equivalent, but certainly the last hundred years of logic suggest that after all it is not very plausible that they should be. It will be instructive to observe a proposed train of reasoning that would show them to be so. This train contains some routine transitions and some interesting ones. Here is what will strike you: None of the interesting transitions concerns the deontic modalities themselves. All are concerned with either the background theory of branching time, or the theory of agency. In the end we shall find out that (7) and (8) are *not* equivalent, though nearly so. And coming to see that sort of thing is part of what logic is for. Marcus, in the shadow of some intolerant and irrational decades that had foolishly put modal considerations on the defensive, said that "establishing the foundations of mathematics is not the only purpose of logic" (Marcus 1960, p. 58). This should count as a tiny example of what she had in mind.

Let us begin with the henscratches for (7) and (8) respectively.

(9) Symbols for (7): $\forall x$*Forbidden:* [α *dstit: P*α*x*].[10]
(10) Symbols for (8): *Forbidden:* [α *dstit:* $\exists x P\alpha x$].

Here is a list of plausible equivalences, which we call the "Main Calculation," that would lead us between (9) and (10). We have indicated on the right how each entry is tied to the one above it. In gently moving quantifiers from the outside to the inside we will run into a number of particular transitions: "*QTF*" means "quantifier-and-truth-function equivalence." Other transitions are applications of previous postulates, as indicated. And the three key transitions "$\forall x/Sett$", "$\forall x/dstit$", and "$\exists x/dstit$" are based on hypothesized quantifier–modality relationships. These three we explicitly discuss below, with an eye to either confirmation or disconfirmation.

Main Calculation

$\forall x \textit{Forbidden:}\ [\alpha\ \textit{dstit:}\ P\alpha x]$	[*Start* = (9)]
$\forall x \textit{Obligated:}\ [\alpha\ \textit{dstit:}\ {\sim}[\alpha\ \textit{dstit:}\ P\alpha x]]$	[Definition 5]
$\forall x \textit{Settled-true:}\ ({\sim}[\alpha\ \textit{dstit:}\ {\sim}[\alpha\ \textit{dstit:}\ P\alpha x]] \supset S_\alpha)$	[Definition 4]
$\textit{Settled-true:}\ \forall x({\sim}[\alpha\ \textit{dstit:}\ {\sim}[\alpha\ \textit{dstit:}\ P\alpha x]] \supset S_\alpha)$	[$\forall x/Sett$]
$\textit{Settled-true:}\ (\exists x\ {\sim}[\alpha\ \textit{dstit:}\ {\sim}[\alpha\ \textit{dstit:}\ P\alpha x]] \supset S_\alpha)$	[*QTF*]
$\textit{Settled-true:}\ ({\sim}\forall x[\alpha\ \textit{dstit:}\ {\sim}[\alpha\ \textit{dstit:}\ P\alpha x]] \supset S_\alpha)$	[*QTF*]
$\textit{Settled-true:}\ ({\sim}[\alpha\ \textit{dstit:}\ \forall x\ {\sim}[\alpha\ \textit{dstit:}\ P\alpha x]] \supset S_\alpha)$	[$\forall x/dstit$]
$\textit{Settled-true:}\ ({\sim}[\alpha\ \textit{dstit:}\ {\sim}\exists x[\alpha\ \textit{dstit:}\ P\alpha x]] \supset S_\alpha)$	[*QTF*]
$\textit{Settled-true:}\ ({\sim}[\alpha\ \textit{dstit:}\ {\sim}[\alpha\ \textit{dstit:}\ \exists x P\alpha x]] \supset S_\alpha)$	[$\exists x/dstit$]
$\textit{Obligated:}\ [\alpha\ \textit{dstit:}\ {\sim}[\alpha\ \textit{dstit:}\ \exists x P\alpha x]]$	[Definition 4]
$\textit{Forbidden:}\ [\alpha\ \textit{dstit:}\ \exists x P\alpha x]$	[*Finish* = (10); Definition 5]

We take up the undiscussed quantifier–modality transitions $\forall x/Sett$, $\forall x/dstit$, and $\exists x/dstit$ in turn. Let us first underline what we said earlier: They have nothing to do with deontic logic per se. They are all either about branching time or else about pure agency. They are about the world and our doings, independently of whether we are getting ready to make some rules or set some standards.

5.1. Barcan Formula for Settled-true:

The first transition does not even concern agency, much less obligation. It is a matter for the general theory of branching time, or perhaps the general theory of highways. What are we to say of the principle $\forall x/Sett$ that permutes the *Settled-true:* modality with a universal quantifier over highways? In symbols:

7. POSTULATE. $\forall x \textit{Settled-true:}\ Qx \leftrightarrow \textit{Settled-true:}\ \forall x Qx$.

Not to put too fine a point on it, we think Postulate 7 is good. Here is why. *Settled-true:* at a moment means, close enough, "true in all histories" through that moment. You can therefore plainly see that *Settled-true:* has the intent of an S5 necessity. That means that the permutation of the universal quantifier with *Settled-true:* is precisely the Barcan formula in one of its guises.[11] Now the Barcan formula is not only time-honored and of exceedingly noble parentage, but of tremendous logical utility. Proof systems without the Barcan formula are awkward and difficult for us to use. Adding the Barcan formula simplifies and thereby increases our logical acumen. It makes us smarter. That's the benefit.

But, you may ask, at what cost? If we use the Barcan formula we must suppose that the range of the quantifier does not vary with the history.[12] Now here is a general point, and we mean to be echoing one made by Marcus[13]: Although English means what it means independently of what we say, our technical quantifiers mean exactly what *we* say they mean. Their meaning is up to us, and it is, as Carnap emphasized, a practical matter. In applying Carnap's advice, it makes a difference whether the quantifiers we are considering are for general use, or for a case at hand.

For general use it seems clearly best to simplify our logical life by interpreting our technical quantifiers – the ones with which we calculate – so that their ranges do not depend on the particular historical continuation.[14] As Marcus argues, for example in Marcus 1961, we shall still be able to find appropriate and indeed illuminating symbolic formulations for every one of those English sentences that might have led us to abandon the Barcan formula, had we been so weak on our philosophical pins as to be bowled over by derisive counterexamples hurled down the alley in an effort to obtain a crowd-pleasing strike. Need we add that such presumed counterexamples will be found on inspection to be made from a nonmetallic composition material, and to contain a hole the size of your thumb?

Proceeding to the local case at hand, observe that the particular example involves quantification over highways. That makes it sound as if in this locality the Barcan formula should fail, since surely what highways there will be depends on what happens and thus on particular historical continuations. Well, yes, the "since" clause is true, but the argument is not. The argument derives from the paradigm of Kripke or Lewis total worlds, each with its worldwide quota of highways. That paradigm, however, is not wanted here. We are looking at a particular concrete moment of forbidden parking. Now whether or not something is a highway at a particular moment does *not* depend on historical continuations of that moment. The domain of highways at a moment (as we might say) is indeed independent of what historically happens next. It is, in the phrase of Castañeda adopted by Bartha 1991, a "circumstance" of the parking. Thus we should accept the permutation labeled "$\forall x$/*Sett*" in general, and therefore in the particular case.

5.2. *Universal Quantifier over dstit*

The next transition we discuss, namely ∀*x*/*dstit*, is also Barcan-like:

(11) General Form: ∀*x*[α *dstit:* *Qx*] ↔ [α *dstit:* ∀*xQx*].
(12) Special case: ∀*x*[α *dstit:* ~[α *dstit:* *P*α*x*]] ↔ [α *dstit:* ∀*x* ~[α *dstit:* *P*α*x*]];

Here, however, appearances deceive: *dstit* is not really a straightforward S5-like necessity modality. As you can see from Definition 2, the semantics for [α *dstit:* *Q*] involves not only a 'positive condition' saying that the (perhaps vacuous) choice that α makes necessitates *Q*, but also a 'negative condition' saying that "it might have been otherwise than *Q*" (from which it follows that the choice was not after all vacuous). We can separate the positive from the negative. For the positive we can adapt the Chellas[15] modality, Δ(α, *Q*), and for the negative we can take the standard branching-time dual of *Settled-true:*, for which we here use *Possibly:*, and sometimes *Can:*.

8. DEFINITION. Δ(α, *Q*) is true at a moment/history pair *m*/*h* just in case *Q* is true at m/h_1 for every history h_1 through *m* that lies with *h* in the same possible choice for α at *m*. (Note the absence of a 'negative condition'.)

9. DEFINITION. *Possibly: Q* ↔ ~*Settled-true:* ~*Q*.

10. DEFINITION. *Can:* [α *dstit:* *Q*] ↔ *Possibly:* [α *dstit*: *Q*].

The rule is that *Possibly:* is a general-purpose modality whose complements are unrestricted, but that *Can:*, although it means quite the same thing, is grammatically restricted to agentives.[16] Then we can read '*Can:* [α *dstit:* *Q*]' as 'α can see to it that *Q*'.

Then we can express [α *dstit:* *Q*] as a conjunction of separate positive and negative conditions:

11. FACT. [α *dstit:* *Q*] ↔ (Δ(α, *Q*) & *Possibly:* ~*Q*).[17]

The advantage is this. The Chellas modality, having only a positive and not a negative condition, is an approximation to agency that is less complex than *dstit*. For it we can transparently see that a Barcan formula is appropriate:

12. POSTULATE. ∀*x*Δ(α, *Qx*) ↔ Δ(α, ∀*xQx*).

The Barcan formula of Postulate 12 is indeed a principle distinctive of the modal logic of agency, but the explaining idea remains the same as for the permutation of the universal quantifier with the *Settled-true*: modality; namely, the domain of quantification – or anyway the domain of highways – at a moment is independent of what happens next, whether by choice or by chance.

If we now chase through the adventures of the universal quantifier in the Special Case (12), it becomes clear that the left and right sides of (12) can, with the important help of both the Barcan formula Postulate 7 and the Barcan formula Postulate 12, be reduced to the following:

(13) Left Side: $\Delta(\alpha, \forall x{\sim}P\alpha x)$ & $\forall x$*Possibly:* $P\alpha x$.

(14) Right Side: $\Delta(\alpha, \forall x{\sim}P\alpha x)$ & $\exists x$*Possibly:* $P\alpha x$.

The left side (13) evidently implies the right side (14), since we have assumed that there are some highways. In contrast, anyone can see that it is not plausible, even in the Special Case, that the Right Side (14) should imply the Left Side (13). Nothing is going to make it plausible that whenever *some* highway is such that α can park on it, it is also true that *each* highway is such that α can park on it. There are just too many highways, many of them remaining in Connecticut even when agent α is about to park in Oregon.

As a consequence of the negative condition *Possibly*:~*Q*, then, we lose the Barcan formula for $\forall x$/*dstit*. It is natural to ask whether we should keep the negative condition at such a high price. We could redefine [α *dstit*: *Q*] as Δ(α, *Q*). Then (12) would be valid; indeed, the entire Main Calculation would go through quite easily. There are, nevertheless, compelling reasons for retaining the negative condition. For one thing, the simplified semantics would make *Obligated*: [α *dstit*: *Q*] valid for every tautology *Q*. More generally, dropping the negative condition would blur the distinction we have made between the evaluative and prescriptive modal families. We intended to limit the discussion to the 'ought to do'. The negative condition is intended to reflect a basic fact about choice: an agent can only see to *Q* if it is possible (at that very moment) for *Q* to be false.

What is the upshot? We return to this after discussing the third and last transition of the Main Calculation.

5.3. Existential Quantifier over dstit

The transition $\exists x$/*dstit* of the Main Calculation is, interestingly, the one that is most bound up with agency itself. It will turn out to be valid, but delicately so: We write a General Form (invalid), a Special Case (also invalid), and then a Special Case in Context (valid).

(15) General Form: $\exists x$[α *dstit:* *Qx*] ↔ [α *dstit:* $\exists x$*Qx*].
(16) Special Case: $\exists x$[α *dstit:* *P*α*x*] ↔ [α *dstit:* $\exists x$*P*α*x*].
(17) Special Case in Context: [α *dstit:* ~$\exists x$[α *dstit:* *P*α*x*]] ↔ [α *dstit:* ~[α *dstit:* $\exists x$*P*α*x*]].

You should see $\exists x/dstit$ as a motion of the existential quantifier between the inside and outside of the *dstit* modality. The General Form is certainly to be rejected in passing from right to left. When in a game of chance one rolls an ordinary pair of dice, one sees to it that they come up some number or other (2 through 12), but it is false that there is a number such that one sees to it that they come up that number. The former is taking one's turn. The latter is cheating.

We have, however, to look at the Special Case. What is most obvious about the Special Case is that for each value of x, the complement of the *dstit* on the Left Side is itself agentive (by Postulate 3). Even with this extra information, however, the equivalence fails. Briefly put, what holds instead is this. Provided $P\alpha x$ is agentive for every value of x,

(18) Modified Special Case: $(\exists x\ [\alpha\ \mathit{dstit}\!: P\alpha x]\ \&\ \mathit{Possibly}\!: \sim\exists x P\alpha x) \leftrightarrow [\alpha\ \mathit{dstit}\!:\ \ \exists x P\alpha x]$.

We might not have noticed this because the extra conjunct seems so overwhelmingly plausible in the highway-parking example; it seems difficult to imagine a moment at which it is absolutely settled that there is a highway on which α parks. It is important to realize, however, that the difficulty of imagination arises from the example, parking on highways, not from agency itself.

We would belabor this subtlety except for the following: Even though the General and the Special Cases fail, the Special Case in Context (17) of the $\exists x/dstit$ transition is in fact logically valid, given Postulate 3. The surrounding context makes the difference. We skip the detailed calculation, which indeed depends heavily on the logic of *dstit*,[18] in favor of a mere record:

13. FACT. If $\forall x(Qx \leftrightarrow [\alpha\ \mathit{dstit}\!: Qx])$ then $[\alpha\ \mathit{dstit}\!: \sim\exists x[\alpha\ \mathit{dstit}\!: Qx]] \leftrightarrow [\alpha\ \mathit{dstit}\!: \sim[\alpha\ \mathit{dstit}\!: \exists x Qx]]$.

5.4. *Upshot for the Main Calculation*

If we put all these considerations together we find the following. The transition from top to bottom of the Main Calculation is secure at every link (given of course the postulates). In the bottom-to-top direction, however, the transition $\forall x/Sett$ is weak, requiring as it does an added and implausible premiss that amounts to this: If α can park on *any* highway then α can park on *any* highway; that is,

$\exists x \mathit{Possibly}\!: P\alpha x \supset \forall x \mathit{Possibly}\!: P\alpha x$.

The upshot of the Main Calculation is that to forbid α to park on each highway (as in (9)) is implicitly to forbid α to park on a highway (as in

(10)), but not conversely. The difference arises only in proper no-can-do cases such as those involving parking on highways in Connecticut when α is in Oregon. This suggests the following: If it is forbidden that α park on a highway (as in (10)), then α is forbidden to park on each highway *on which* α *can park*. The suggestion can be verified; (10)) implies the following:

(19) $\forall x(\mathit{Can}{:}\ P\alpha x \supset \mathit{Forbidden}{:}\ [\alpha\ \mathit{dstit}{:}\ P\alpha x])$: α is forbidden to park on each highway on which α can park.

This implication is not reversible. What happens, however, if we adopt a rule according to which

(20) $\mathit{Can}{:}\ [\alpha\ \mathit{dstit}{:}\ \exists x P\alpha x] \supset \mathit{Forbidden}{:}\ [\alpha\ \mathit{dstit}{:}\ \exists x P\alpha x]$: α is forbidden to park on a highway if α can park on a highway.

Here as it turns out we have an even weaker statement, implied by but not implying (19). As a visual help in sorting this out, here is a definition that highlights a form common to (19) and (20).

14. DEFINITION. *Forbidden-if-can-do:* $[\alpha\ \mathit{dstit}{:}\ Q] \leftrightarrow (\mathit{Can}{:}\ [\alpha\ \mathit{dstit}{:}\ Q] \supset \mathit{Forbidden}{:}\ [\alpha\ \mathit{dstit}{:}\ Q])$.

(This definition evidently only makes sense for restricted complements.) Then (19) and (20) are more transparently written in terms of weak prohibition respectively as follows:

(21) $\forall x \mathit{Forbidden\text{-}if\text{-}can\text{-}do}{:}\ [\alpha\ \mathit{dstit}{:}\ P\alpha x]$
(22) $\mathit{Forbidden\text{-}if\text{-}can\text{-}do}{:}\ [\alpha\ \mathit{dstit}{:}\ \exists x P\alpha x]$

So now we have four prohibitions, all subtly different, with irreversible implications as follows:

(9) $\forall x \mathit{Forbidden}{:}\ [\alpha\ \mathit{dstit}{:}\ P\alpha x]$.
(10) $\mathit{Forbidden}{:}\ [\alpha\ \mathit{dstit}{:}\ \exists x P\alpha x]$.
(21) $\forall x \mathit{Forbidden\text{-}if\text{-}can\text{-}do}{:}\ [\alpha\ \mathit{dstit}{:}\ P\alpha x]$.
(22) $\mathit{Forbidden\text{-}if\text{-}can\text{-}do}{:}\ [\alpha\ \mathit{dstit}{:}\ \exists x P\alpha x]$.

Inspection of that array seems enough both to explain and *to correct* our feeling of equivalence between (7) and (8).

Which of these is wanted when we are writing up the traffic regulations for the State of Oregon? The following equivalences should guide us in the task:

(23) (9) $\leftrightarrow$ $[(10)\ \&\ (\exists x \sim \mathit{Can}{:}\ P\alpha x \supset \mathit{Settled\text{-}true}{:}\ S_\alpha)]$
(24) (10) $\leftrightarrow$ $[(21)\ \&\ ({\sim}\exists x \mathit{Can}{:}\ P\alpha x \supset \mathit{Settled\text{-}true}{:}\ S_\alpha)]$
(25) (21) $\leftrightarrow$ $[(22)\ \&\ (\mathit{Settled\text{-}true}{:}\ \exists x P\alpha x \supset \mathit{Settled\text{-}true}{:}\ S_\alpha)]$

These equivalences suggest that (9) and (10) are unacceptable interpretations of the parking prohibition (3). By (23), (9) entails that if there is any highway that α can't park on, then α has no good choice. Given the abundance of highways, (9) would perpetually put all agents (including the lawmakers themselves!) in the no-good-choice situation. Similarly, as (24) shows, (10) entails that if there is no highway that α can park on (for example, when α is sitting at home), then α has no good choice. Neither of these objectionable consequences follows from (21) or (22). The only difference between these two versions of the parking prohibition occurs in the (rare) case of a forced choice. For example, suppose you are moving toward a Y-split between two highways when your engine stops. You are forced to choose between parking on one or the other highway in the strongest possible sense: If you don't park on one you cannot escape parking on the other. In this case, since it is settled true that there is a highway on which you park, we can see from (25), that (21) implies that you suffer the sanction no matter what you do. You are in a "legal dilemma." In contrast, (22) does not provide a sanction in the forced choice described – even if you yourself purposely stopped your engine by turning the key.

It is plausible that the lawmakers in Oregon would want one of the weaker prohibitions, (21) or (22), but less plausible that they would want the weakest. For the rest of this essay we will assume that (21) is the most natural choice.

6. Generalization on Agents

So far there has been talk only of a single agent, α. There is very little that is problematic in adding a generalization to all agents, so that we may say that 'Parking on highways is forbidden' comes to

(26) ∀α∀*xForbidden-if-can-do:* [α *dstit: P*α*x*].

Although this is straightforward, while our attention is concentrated on the agent position, it will be good to raise the question as to its "extensionality." In addressing this question, we want to consider two theses and one piece of advice from Marcus.

Marcus has chiefly discussed when a *system* is extensional, but here we are asking whether a particular *context* [α *dstit: Q*] is extensional in the position α. The questions are of course related, and the first thesis of Marcus is exactly on target, namely, that extensionality always has to do with a plurality of relations 'eq' of equality some of which have stronger replacement principles than others. In the theory of branching-time-with-choices there is enormous room for such equivalence relations, for we may consider with profit *at least* equality relations indexed to a single moment/history pair, or indexed to a single moment with the histories quantified out, or indexed to a single history with the moments quantified out, or de-indexed by double quantification to give the strongest equality relation of all. This seems to be exactly the sort of situation that Marcus is picturing.

The second thesis is this: Marcus says that "extensionality principles are absent on the level of individuals" (Marcus 1961, p. 305). In context her meaning is that plural equalities are available only for sentential or predicative parts of speech.[19] For individuals there is only a single equality predicate, the strongest, that she identifies with identity. She says that weaker (e.g., contingent) equivalences are "*misleadingly* called 'true identities' " (Marcus 1961, p. 312; our emphasis). If her second thesis is right, we have no sensible question. And indeed it surely is correct for the powerful and interesting languages that Marcus has investigated. There are, however, the at least equally powerful and equally interesting languages championed by Carnap and fully developed by Bressan and Montague. In these languages singular terms carry "individual concepts" as their intensional semantic value. These are good and useful languages. Quine has made delicious fun of them over the years, and Kripke and others have cooked up their own pungent mockeries, but surely it is high time to stop dining on proofs by spoof when these are pressed into the service of intolerance.[20] As Pooh would say, such proofs are not sustaining. Here we are led to serve up a particularly satisfying portion of advice from Marcus:

> the polemics of modal logic are perhaps best carried out in terms of some explicit semantical construction. (Marcus 1961, p. 319)

Without, however, fully digesting this advice, first let us say that such constructions, namely those of Bressan and Montague, are standing ready in the kitchen, and second let us abandon the metaphor and just wing it for a while.

Our point is that there is something fundamentally nonextensional about the agent position of *dstit* statements. Even for the single evaluation of a single statement [α *dstit: Q*] at a single moment/history pair, we need to fix the agent at that moment independently of histories. We need to be able to say what choices are available to the agent regardless of what happens next. Furthermore, when we proceed, as we must, to speak of strategies or other concepts that take us from moment to moment, we shall need to fix the concept of the agent in an even more severe sense. All of this is carried by saying that the underlying concept of 'agent' must be 'absolute' in Bressan's sense. This is a great deal like saying, in the language of Marcus 1961 (p. 304) that agents as we need them are "constant objects of reference." There are, however, two differences. First, the Marcus phrase is informal and metalinguistic, whereas Bressan's account is carried by a rigorous object-language but second-order definition of the absoluteness of concepts. Second, the Marcus approach, which in this respect is the same as Kripke's, renders this kind of idea a presupposition of applying quantified modal logic. In contrast Bressan leaves it open for us to make it a matter of extralogical theory – for example, a matter of serious empirical physics in the theory of general relativity – whether a concept is or is not absolute.

There is, however, another side to the point. As we can learn from Bressan, even the most nonextensional of concepts have extensional cousins that arise by "extensionalization" with suitably weak equality relations. Given any Qx however fierce, there is always $\exists x_1((x \text{ eq } x_1) \,\&\, Qx_1)$. The predicate of parking on a highway, and *dstit* in general, are like that. All we want to add to these unfortunately cryptic remarks is this: Postulate 3 and the various definitions are intended to hold only for α that fall under the absolute concept of agent, and not for extensional equivalents thereof.

This leaves much that is problematic for us to treat on some other occasion, perhaps the same occasion on which we can discuss what is needed to cater to the natural wish to restrict the range of "α" in (26) to some sane set of agents.

7. Temporal Generalization

As note 4 said, this development has suppressed explicit reference to "time" in favor of locutions that support "truth at a moment/history pair," where the moment/history is supplied externally (either indexically or by some implicit binding). This linguistic policy simplifies some considerations, but if we wish to frame prohibitions that generalize on moments in some carefully controlled way, then we should need to enrich the language. The easiest and most transparent thing to do would be to add variables explicitly ranging over moments or histories or moment/history pairs, depending on what is convenient.[21]

Since our purpose here is merely illustrative, however, we refrain from such additions. We suppose instead that we only require the language to describe a prohibition good throughout the entire future, regardless of what happens. For that all we need is the standard tense-logical "will-always" construction:

15. DEFINITION. *Will-always*: Q is true at a moment/history pair m/h iff Q is true at m_1/h for every moment m_1 belonging to h that is properly later than m.

Thus, *Settled-true: Will-always: Q* (sometimes written as *Inevitable: Q*) is true at m/h just in case Q is settled true at each moment that is properly later than m.

We are now in a position to offer a "final" account of (3) as a generalized prohibition of parking on highways binding at all future moments no matter what happens:

(27) *Settled-true: Will-always:* ∀α∀x*Forbidden-if-can-do:* [α *dstit: P*α*x*]: Parking on highways is henceforth forbidden.

We hope our insertion of "henceforth" is self-explanatory.

8. The Outer Ought

What, finally, of the outer deontic modality of (1)? Marcus 1966 says the 'ought' in (1), 'Parking on highways ought to be forbidden,' should be taken to be evaluative. That is plausible, but we nevertheless want to run with the prescriptive possibility, just to indicate what is not so obvious, namely, that it makes perfectly good sense. Permit us therefore to take as a temporary target

(28) Parking on highways ought to be forbidden *by someone*,

or, in the active voice,

(29) *Someone* ought to forbid parking on highways.

The critical and much-overlooked point is this. 'To forbid' is itself very naturally construed as an agentive verb. If we were pursuing ontology we would have to force a search for the 'action' of forbidding each agent to park on a highway. Fortunately, however, no such wild ontological chase is called for. The thing is that we *already* have what is needed to make entirely good modal sense out of one agent β forbidding another agent α, or as in the present example, forbidding all agents. To forbid an agent α to park is to see to it that α is forbidden to park. To forbid all agents to park is to see to it that all agents are forbidden to park. No tricks, just an appropriate *dstit*.

(30) [β *dstit: Settled-true: Will-always:* ∀α∀*xForbidden-if-can-do:* [α *dstit: P*α*x*]]

Now to say that β *ought* to forbid parking on highways is just to use language already at our disposal:

(31) *Obligated:* [β *dstit: Settled-true: Will-always:* ∀α∀*xForbidden-if-can-do:* [α *dstit: P*α*x*]]

What goes wrong at this point is that we don't much feel like existentially generalizing this to say that there is some particular agent that ought to do the forbidding. And the Restricted Complement Thesis prevents us from putting that existential quantifier *inside* the *Obligated:* modality (though we could perfectly well do that if the modality were impersonal).

There is, however, a natural and proper way in which to express the generality of the not-very-personal 'ought' while still treating the 'ought' as prescriptive and while still hewing to the Restricted Complement Thesis. Instead of 'someone' in the logicians' sense conveying quantification over agents, it would be natural to say,

(32) *They* [or ''*we*''] ought to forbid parking on highways.

Here the subject 'they' of the English sentence is intended not as a singular quantifier phrase, but as the name of a class or aggregate in the sense elaborated by Marcus in Marcus 1963 and Marcus 1974. The obligation is collective, not distributive. If there were no theory about such constructions this would be just throwing sand in your eyes, but in fact there is. Take Γ as some suitable aggregate of agents, and use the theory of *joint* agency from Belnap and Perloff 1991. Then (32) is well symbolized by the following.

(33) *Obligated:* [Γ *dstit: Settled-true: Will-always:* $\forall\alpha\forall x$*Forbidden-if-can-do:* [α *dstit:* $P\alpha x$]]

This says that there is a *joint* obligation on Γ taken collectively to see to it that parking on highways is henceforth forbidden. If we apply the Anderson/*dstit* simplification of Section 3 to this extended form of obligation, we shall find that we need to index sanctions not only by single agents α, as in $\boldsymbol{S}_\alpha$, but also by groups of agents: $\boldsymbol{S}_\Gamma$. That is of course just a formal remark. The concrete content stands in need of exploration. Our progress here is limited to seeing the difference between (33) on the one hand and, on the other, either

(34) $\exists\beta$*Obligated:* [β *dstit: Settled-true: Will-always:* $\forall\alpha\forall x$*Forbidden-if-can-do:* [α *dstit:* $P\alpha x$]],

which doesn't mean what we wish, or

(35) *Obligated:* $\exists\beta$[β *dstit: Settled-true: Will-always:* $\forall\alpha\forall x$*Forbidden-if-can-do:* [α *dstit:* $P\alpha x$]]

which is not only unwanted but is also a disagreeable violation of the grammar required by the Restricted Complement Thesis.

One last complication. The point of view of modality makes us see very clearly that by one more degree of nesting we could express that whereas (i) the *obligation* is joint on the group, nevertheless (ii) the particular *dstit* is to be carried out by some member of the group. In this case the *dstit* is of a forbidding, so that we have the following:

Obligated: [Γ *dstit:* $\exists\beta((\beta \in \Gamma)$ & [β *dstit: Settled-true: Will-always:* $\forall\alpha\ \forall x$ *Forbidden-if-can-do:* [α *dstit:* $P\alpha x$]])]

Again the thing to notice is that the new subtlety requires no tricks. Nothing is made up to order. All is transparent. We are just using *the power of modalities combined with quantifiers* to which we have become entitled by the philosophical pioneering of Ruth Barcan Marcus.

NOTES

1. We are indebted to M. Green for suggesting revisions both small and great.
2. This version is presented in various papers by Perloff and Belnap jointly and separately. Belnap and Perloff 1988, which is earliest, gives a suitable introduction. Belnap and Perloff 1992, which is most recent, lists the intervening papers.
3. *Stit* and *dstit* have much in common, but to say more is to say too much. Aside from occasional squibs below, it is necessary to refer the interested reader to the papers mentioned in note 2.
4. We are leaving the '*t*' in Marcus's '*x* does *w* at *t*' to be indicated indexically rather than explicitly, but it is nevertheless there. That is, (6) speaks of a particular occasion of parking on highway *x*, not a habit or practice. Also we are using α as ranging over agents, because that is what the rest of the *stit* literature does, and we are using *x* as ranging over highways for convenience.
5. Anderson introduced the 'sanction' in 1956. There is not the slightest implication that this is the best that can or should be done in the way of representing norms. Advances since Anderson 1956 are legion. Let us only note the interdisciplinary study Horty 1991, since Horty was also one of those responsible for *dstit*.
6. Quine tells us that only truth functions can be understood. Dummett tells us that no truth functions can be understood. Any schoolman can see, however, that these exotic positions are contraries: They are capable of simultaneous falsehood.
7. The effort to straighten us out on this goes back at least to Hohfeld 1919; see esp. Makinson 1986.
8. This is a point at which *dstit* differs from *stit*.
9. Or each highway within a given jurisdiction. We shall suppress this subtlety.
10. It would be acceptable to shorten [α *dstit: Pαx*] to *Pαx*, just because the latter is agentive, i.e., they are equivalent. But it is probably better to use the longer 'normal form' in order to remind us that the complement of *Forbidden:* must always be a *dstit* or something equivalent. This is in accord with the Restricted Complement Thesis mentioned in the text. Let us add here that by the Restricted Complement Thesis a third candidate, *Forbidden:* ∃*x*[α *dstit: Pαx*], is not on the face of it grammatical. We should suspect that this third proposed complement of *Forbidden:* might, in the language of Marcus 1966, describe a mere state of affairs that could not meaningfully be on α's list of prohibitions. And in fact the analysis of the ∃*x*/*dstit* transition subsequently in the text confirms this suspicion.
11. From Barcan 1946. So-named by Prior 1957.
12. So much is clear from the Kripke analysis.
13. For example, in Marcus 1972.
14. This still leaves numerous options, of which the following are salient: (i) Marcus quantifiers ranging over actualia; (ii) quantifiers ranging over possibilia; (iii) Bressan 'individual concept' quantifiers with intensional predication; and (iv) substitutional quantifiers. Arguments that some alternative is better or worse than one of these have tended to be based on mockery involving the double modal, 'How could you say possibly such a thing?' That's amusing, but also sad. For many practical purposes of *serious* formalization, it is appropriate to use the much-neglected Bressan individual concept semantics with intensional predication, of which we say just a little more below.
15. See Chellas 1969 and 1992.
16. The parallel analysis for *stit* must go differently. There is a much more marked contrast between the 'can-be' and the 'can-do' for *stit*.

17. Chellas 1992 has emphasized this. As he notes, the same disentanglement is not possible for *stit*.
18. The analysis for *stit* would have to be substantially different.
19. Marcus 1961, p. 304, makes a passing reference to "similarity or congruence," which are not here at issue. Also not germane is her discussion of the identity of indiscernibles as a kind of extensionality principle for individuals.
20. In contrast it seems to us quite in order to ridicule positions that are themselves intolerant. Sometimes that is the only way to neutralize some prejudice that stands blocking the road to inquiry.
21. Some people think doing so would be to import 'the metalanguage' into 'the object language'. That opinion, which should by now be a relic, is demonstrably false; as Bressan 1972 has in effect shown, this linguistic power is not only useful, but already *definable*.

REFERENCES

Anderson, A. R. The Formal Analysis of Normative Systems. Technical report, no. 2, Contract no. SAR-NONR-609(16), Office of Naval Research, New Haven, Conn., 1956. Reprinted in N. Rescher (ed.), *The Logic of Decision and Action* (Pittsburgh: University of Pittsburgh Press, 1967), pp. 147–213.

Barcan, R. 1946. A Functional Calculus of First Order Based on Strict Implication. *Journal of Symbolic Logic* 11: 1–16. See also Marcus.

Bartha, P. 1991. Conditional Obligation and the Logic of Agency. In Meyer and Wieringa 1991, pp. 89–106. Forthcoming in *Annals of Mathematics and Artificial Intelligence (AMAI).*

Belnap, N., and M. Perloff. 1988. Seeing to It That: A Canonical Form for Agentives. *Theoria* 54: 175–99. Corrected version in H. E. Kyburg, Jr., R. P. Loui, and G. N. Carlson (eds.), *Knowledge Representation and Defeasible Reasoning*. Studies in Cognitive Systems, vol. 5. Dordrecht/Boston/London: Kluwer Academic Publishers, 1990, pp. 167–90.

1991. In the Realm of Agents. In Meyer and Wieringa 1991, pp. 107–26. Forthcoming in *Annals of Mathematics and Artificial Intelligence* (*AMAI*).

1992. The Way of the Agent. *Studia Logica* 51: 463–84.

Bressan, A. 1972. *A General Interpreted Modal Calculus*. New Haven, Conn.: Yale University Press.

Chellas, B. F. 1969. *The Logical Form of Imperatives*. Stanford, Calif.: Perry Lane Press.

1992. Time and Modality in the Logic of Agency. *Studia Logica* 51: 485–518.

Gabbay, D., and G. Guenthner, eds. *Handbook of Philosophical Logic,* Vol. II: *Extensions of Classical Logic*. Synthèse Library, Studies in Epistemology, vol. 165. Dordrecht: D. Reidel, 1984.

Hohfeld, W. N. 1919. *Fundamental Legal Conceptions as Applied in Judicial Reasoning, and Other Essays*. New Haven, Conn.: Yale University Press.

Horty, J. F. 1989. "An Alternative *stit* operator." Unpublished manuscript, Department of Philosophy, University of Maryland.

Horty, J. 1991. Moral Dilemmas and Nonmonotonic Logic. Unpublished manuscript, 1991. Forthcoming in *Journal of Philosophical Logic*.

Makinson, D. 1986. On the Formal Representation of Rights Relations. *Journal of Philosophical Logic* 15: 403–25.

Marcus, R. B. 1960. Extensionality. *Mind* 69: 55–62.

1961. Modalities and Intensional Languages. *Synthèse* 13: 303–22.

1963. Classes and Attributes in Extended Modal Systems. *Acta Philosophica Fennica* 16: 123–36.

1966. Iterated Deontic Modalities. *Mind* 75: 580–2.

1972. Quantification and Ontology. *Noûs*, 6: 240–50.

1974. Classes, Collections, and Individuals. *American Philosophical Quarterly* 11: 227–32.

1977. Dispensing with Possibilia. In *Proceedings of the American Philosophical Association*, pp. 39–51.

1980. Moral Dilemmas and Consistency. *Journal of Philosophy* 77: 121–36.

1985–6. Possibilia and Possible Worlds. *Grazer Philosophische Studien* 25, 26: 107–33.

Meyer, J.-J. Ch., and R. J. Wieringa, (eds.) 1991. *△EON '91: First International Workshop on Deontic Logic in Computer Science*. Amsterdam, December 11–13.

Prior, A. 1957. *Time and Modality*. Oxford: Oxford University Press.

Thomason, R. H. 1984. Combinations of Tense and Modality. In Gabbay and Guenthner 1984, pp. 135–65.

von Kutschera, F. 1986. ''Bewirken.'' *Erkenntnis* 24: 253–381.

III

Belief

12

Pierre, Saul, Ruth, and Bob and a Puzzle about Belief

ROBERT J. FOGELIN

Pierre

In "A Puzzle about Belief," Kripke tells the following story about Pierre:

On the basis of what he has heard of London, [Pierre] is inclined to think that it is pretty. So he says, in French, "*Londres est jolie.*" On the basis of his sincere French utterance, we will conclude:
(4) Pierre believes that London is pretty.
[Finding himself in a somewhat squalid part of London, and having learned English by the "direct method" (i.e., without a translation manual), Pierre notices that his] surroundings are . . . unattractive, and he is unimpressed with most of the rest of what he happens to see. So he is inclined to assent to the English sentence:
(5) London is not pretty.
He has *no* inclination to assent to:
(6) London is pretty.
. . . [O]n the basis of his sincere assent to (5), we should conclude:
(7) Pierre believes that London is not pretty. (254–5)[1]

Pierre, as the story is told, does not recognize that 'Londres' and 'London' are names for the selfsame city.

Saul

It is important to see that there is nothing inherently puzzling about this narrative. The events, though exotic, could have unfolded just as Kripke describes them. Furthermore, nothing prevents us from saying that Pierre holds inconsistent beliefs concerning London: *Of* it, he holds beliefs that cannot both be true. Indeed, the story tells us how this came about.

A puzzle is supposed to arise, however, when certain seemingly acceptable principles are applied to this situation. One such principle Kripke calls the *disquotational principle:*

If a normal English speaker, on reflection, sincerely assents to '*p*,' then he believes that *p*. (248–9)

Kripke also makes use of what he calls a "strengthened 'biconditional' form of the disquotational principle":

A normal English speaker who is not reticent will be disposed to sincere reflective assent to 'p' if and only if he believes that p. (249)

Kripke also invokes what he calls the *principle of translation*:

If a sentence of one language expresses a truth in that language, then any translation of it into any other language also expresses a truth (in that other language). (250)

I take it that Kripke has in mind a *correct* translation. Finally, we need an additional factual premise:

'London is pretty' is the correct translation of 'Londres est jolie'.

From all this, we seem able to conclude:

(4_{dd}) Pierre holds the belief: London is pretty.
(7_{dd}) Pierre holds the belief: London is not pretty.

Here the subscripts indicate that Kripke is giving a so-called *de dicto* (dd) reading of these remarks about Pierre's beliefs. This reading contrasts with a so-called *de re* (dr) reading that might be phrased as follows:

(4_{dr}) Of London, Pierre believes it is pretty.
(7_{dr}) Of London, Pierre believes it is not pretty.

This pair of *de re* constructions may or may not be problematic. Kripke does not take an explicit stand on this matter. They are noted here to underscore Kripke's claim that he is not concerned with them but, instead, is concerned only with *de dicto* reports of Pierre's beliefs. Later we will see that Ruth Marcus challenges this *de dicto* interpretation of Pierre's beliefs and, in doing so, shifts the grounds of the discussion in a fundamental way.

Finally, given this *de dicto* reading, it seems possible to reach a flat-out contradiction if we adopt the stronger, biconditional, version of the disquotational principle:

A normal English speaker who is not reticent will be disposed to sincere reflective assent to '*p*' if and only if he believes that *p*. (249)

That principle, together with premises already given, allows us to conclude that Pierre holds the belief that London is pretty and also does not hold this belief. This, Kripke tells us, is a "contradiction, not merely in Pierre's judgments, but in our own" (258).[2]

There are puzzles and there are puzzles: why does Kripke think that we should take this one seriously? He has two reasons. His first reason is systematic: Kripke thinks that this really *is* a puzzle, and a large part of his paper is dedicated to showing that no solution he can find is satisfactory. He himself offers no solution to it. His second reason is dialectical: It concerns opposing theories of the meaning of proper names. Broadly, and very roughly, there have been two main rivals for the account of how proper names function. The older theory – associated with Locke and Mill, and revived by Marcus, Kripke, Donnellan, and others – is that the meaning of a proper name is the object it refers to, or, perhaps, that the sole linguistic function of a proper name is to refer to something. The newer theory – found in the writings of Frege and modified by Russell, Wittgenstein, Searle, and others – is that proper names refer in virtue of having a descriptive content, what Frege called a *sense*. I'll call the first sort of theory a pure reference theory of proper names, the second a sense theory of proper names.

To go over familiar ground quickly (and not carefully), the sense theory of proper names first recommended itself to Frege, and later to others, because it seemed to provide a way of dealing with a series of otherwise intractable problems. One concerned the significance of identity claims where the sign for identity is flanked by proper names (''Phosphorous is Hesperus''). Another concerned the failure of the substitutivity of coreferential names in opaque contexts (''Ellen believes that Cicero wrote *De Finibus*, but she does not believe that Tully wrote *De Finibus*''). Another concerns claims that someone does not exist where the referring term is a proper name (''Prester John did not exist''). There are other problems about proper names, but this is enough. The initial attraction of the sense theory of proper names is that it seemed to offer a systematically motivated answer to all these problems. The pure reference theory seemed unable to solve any of them.

Those who have attempted to revive the pure reference theory of names thus start out with a double deficit. They will have to show that a pure reference theory can, despite Fregean arguments to the contrary, give an adequate account of proper names in such contexts. They will also have to show that a pure reference theory can deal with these problems at least as well as the *best* variant of a sense theory. In *Naming and Necessity* Kripke made some apparent progress along these lines with respect to the identity problem.[3] His key move was to distinguish what he called the *metaphysical* notion of necessity from the *epistemological* notion of the a priori. He then argued that identity statements, when true, are necessarily true but, for all that, can be known to be true a posteriori. He also argued in detail that no available version of a sense theory – despite the initial attraction of the approach – is adequate to the task. Kripke did not, however, attempt to extend this double strategy to the two other standard problems concerning proper names, and they have, in fact, proved stubbornly recalcitrant to solution within a pure reference theory.

In "A Puzzle about Belief," Kripke adopts a new strategy for handling the confrontation between the sense theory and the pure reference theory. He claims to have discovered a deep problem – perhaps even a paradox – in the occurrence of proper names in belief contexts. He then elevates the solution of this problem to the status of a success condition on any adequate account of proper names. By what right he elevates this particular problem to this status rather than, say, the problem of making sense out of claims like 'Prester John did not exist,' he does not say. The move calls to mind Kant's famous remark in his *Prolegomena*:

> All metaphysicians are . . . solemnly and legally suspended from their occupations till they shall have adequately answered the question "How are synthetic cognitions *a priori* possible?"[4]

Unlike Kant, however, Kripke gives no answer to his own question but asserts instead that none of the competing theories is able to answer it either. By this maneuver Kripke tries to level a playing field that initially seemed tilted in favor of sense theories.

The dialectical situation is this. Kripke claims to have found a puzzle concerning the occurrence of proper names in belief contexts that no contemporary theory of reference can solve. Against this, I will argue that the puzzle concerns the reference of proper names only incidentally. Furthermore, setting aside *independent* difficulties, as far as I can see any moderately plausible theory of reference, including pure reference theories and sense theories (in their variations and hybrid combinations), will have no difficulty solving Kripke's puzzle, simply because it is not much of a puzzle. I think that there are unanswered objections to all the theories of reference currently available, but the ability to deal with Kripke's puzzle is not among them. With respect to that puzzle – setting aside independent difficulties – they all come out winners. This is a result that Kripke would not welcome.

Bob

Consider an example that parallels Kripke's.[5] When intoxicated, Martin believes himself to be a great lover. To be more precise, when drunk, he believes himself to be a great lover both drunk and sober. If someone asks the drunken Martin if he is a great lover, he will wink and nudge and say with bravado, "Are you kidding?". If the same question is asked of the sober Martin, he sighs and wanly says, "Are you kidding?". We will further suppose that Martin, when sober, never reflects upon the opinions he has when drunk and, similarly, when drunk is unconcerned with his opinions when sober. He has no recognition of his shifting views on this matter. What about Martin, does he believe that he is a great lover or not? The natural answer is that he believes it when drunk, and doesn't (indeed, even disbelieves it) when he is sober. It seems, then, that Martin has contradictory beliefs that get triggered under different circumstances.

There are, of course, ways of avoiding the attribution of contradictory beliefs to Martin. We might say that there really are two Martins, Martin_d and Martin_s, where the first Martin believes that he is a great lover and the second does not. Unless Martin undergoes a violent personality change when drunk – or genuinely does have multiple personalities – this seems a wholly desperate maneuver. We could also adopt the Bacchic Principle of Disquotation: *in vino veritas*. It might be formulated in this way:

Whatever a normal speaker sincerely assents to when drunk, he *really* believes, i.e., believes both drunk and sober.

In this way we can avoid attributing contradictory beliefs to a single person, but, of course, the Bacchic Principle seems plainly false. We could also maintain that a person does not *really* believe things that he assents to when drunk if he does not also assent to them when sober, but that seems equally arbitrary. It seems much more natural, and wholly unproblematic, to say that someone can believe something in certain settings or under certain conditions, but not believe it in other circumstances or in other settings. Let us say that such a person has a divided belief system – or simply, divided beliefs. A person has a divided belief system if he believes something in one subsystem that he does not believe in another. I will say that a person's belief system is strongly divided if that person believes something in one system and disbelieves it in another. As described, Martin has a strongly divided belief system relative to his sexual prowess.[6]

Clearly, if someone insists that we give a straightforward answer to the question, "Does Martin believe that he is a great lover?," we have a right to refuse to answer this question directly. We have a right to say instead that it all depends on whether he is drunk or sober. So we have reached the following modest conclusion: Questions of the form 'Does *S* believe *p* or not?' may not be directly answerable – that is, answerable with a simple yes or no – if *S* has divided beliefs relative to *p*. Note – and this is important – that this is a perfectly general principle concerning belief attribution and has nothing specifically to do with language. We sometimes attribute beliefs to people on the basis of their nonlinguistic behavior, and differing behavior in different contexts can be a sign of divided beliefs.

Back to Pierre: Does he believe that London is pretty? Tutored by our previous example, we can say that he believes it in his French-thinking mode and does not believe it in his English-thinking mode. This is not an arbitrary description of the situation, for it simply reports the facts as Kripke presented them. Furthermore, if we are pressed to say whether Pierre really believes that London is pretty or believes that it is not, we have a right to refuse to give a simple yes or no answer this question, just as we have a right to refuse to give a simple yes or no answer concerning Martin's beliefs.

This last point is central to my disagreement with Kripke, for his puzzle is driven by his insistence that we *do* have to give a simple yes/no answer to the question concerning Pierre's beliefs, and that not to do so is an evasion. Kripke insists on this a number of times, perhaps most forcefully in this passage:

> Does Pierre, or does he not, believe that London is pretty? I know of no answer to *this* question that seems satisfactory. It is no answer to protest that, in some *other* terminology, one can state 'all the relevant facts'. . . . As in the case of logical paradoxes, the present puzzle presents us with a problem of customarily accepted principles and a challenge to formulate an acceptable set of principles that does not lead to paradox, is intuitively sound, and supports inferences we usually make. (259)

To take Kripke's challenge head on, let us look again at his disquotational principle:

> If a normal English speaker, on reflection, sincerely assents to '*p*', then he believes that *p*.

If what has been said so far is correct, this principle stands in need of emendation. We attribute beliefs to people on various grounds, for example, by observing their purposive behavior, through knowing what most people believe, and also through monitoring their sincere assents. On whatever basis we make such an attribution, we always have to add the proviso that the person is not a divided believer with respect to the belief in question. So the disquotational principle must be rewritten as follows:

> If a normal English speaker, on reflection, sincerely assents to '*p*', then he believes that *p*, provided the speaker is not a divided believer with respect to *p*.

Normally, of course, we are right in assuming that we are not dealing with divided believers, so we may safely ignore the proviso concerning them. When, however, we are dealing with divided believers like Pierre and Martin, the proviso obviously has its place. With its introduction, the inference from sincere reflective assent to belief is blocked, and Kripke's problem is solved. I realize that this answer trivializes Kripke's puzzle, but that is my intention.

To see the force of this response to Kripke's puzzle, we can note that it allows us to give straightforward answers to questions that Kripke finds unanswerable or at least problematic. They concern Pierre's beliefs at various stages of his career. Kripke does not think that it is possible to give consistent answers to all of them. I do. First, did Pierre believe that London was pretty before he went to London? If the story as told is relevantly complete, that is, if there are no other reasons for believing that Pierre's belief system is aberrant, the answer is yes. Before going to London, Pierre was not a divided believer. After he goes to London and acquires English, and so on, does he then believe that London is pretty? Since he is now a divided believer, there is no direct answer to that question. After he has been to London and acquired English and so on, should Pierre "*retroactively* be judged *never* to have believed that London is pretty" (256)? Of course not. Finally, if "an electric shock wiped out all of his memories of the French language, what he learned in France, and his French past" (257), would he then (as the story is told) believe that London is not pretty? Yes! Whatever oddness may seem to attach to these responses is fully explained by the exotic character of the story itself.

In trying to convince us that the answers to these questions are problematic, indeed, deeply problematic, Kripke says a number of things that will not bear scrutiny. One is this:

> Should we say that Pierre, now that he lives in London and speaks English, no longer believes that London is pretty? Well, unquestionably Pierre *once* believed that London is pretty. So we would be forced to say that Pierre has *changed his mind, has given up his previous belief.* (256)

Of course, we are not constrained to say anything of the sort. Becoming a divided believer need not involve changing one's mind or giving something up. Indeed, mind-changing and giving up a belief are ways to avoid becoming a divided believer. Pierre became a divided believer simply through acquiring new beliefs.

A second curious feature of Kripke's discussion is the mileage he attempts to make out of the fact that "as an English speaker, he does not differ at all from them [i.e., the monolingual Anglophones who surround him]" (256). He does, of course, differ from them in a respect that lies at the heart of Kripke's puzzle: He is a divided believer with respect to the prettiness of London, while they, presumably, are not.

At various places I have remarked that there is no direct way of answering the question "Does Pierre believe that London is pretty?" after he has gone to London. We can, as Kripke concedes, give an adequate *description* of Pierre's doxastic situation; indeed, as he further concedes, such a description can be presented in a way that generates nothing paradoxical. We get a paradox – or at least a puzzle – only if we are forced to give a direct answer to the question "Does Pierre believe that London is pretty?" We have seen that the disquotational principle properly modified puts us under no such constraint. We might, however, wish to introduce a weaker version of the disquotational principle to allow relativized ascriptions of beliefs to Pierre. Just as we might distinguish what Martin believes when sober from what he believes when drunk, or what he believes in church from what he believes in the laboratory, and so forth, we might distinguish what Pierre believes in his Francophone mode from what he believes in his Anglophone mode. In line with this, we might construct a relativized version of the disquotational principle that would look something like this:

If a normal English speaker, on reflection, sincerely assents to '*p*', then he believes that *p* relative to the belief system from within which this assent is given.[7]

If the person is not a divided believer, it is safe to ignore relativization to a particular system of beliefs, for the relativization will be to the entire system of belief and thus be noncontrastive. But however the details are worked out, this relativized version of the disquotational principle, like the restricted version I discussed before, does not yield a problem – not to say a paradox – when combined with the translation principle and the fact of Pierre's reflective and sincere assent to various sentences concerning the prettiness of London.

So far I have been arguing against Kripke's claim that his puzzle *is* a puzzle. I now want to turn to a second matter briefly mentioned before, namely, that the puzzle has very little (if any) bearing upon contemporary conflicts in the philosophy of language. To repeat a general point already made, any principle for the ascription of beliefs, whether it has a linguistic or nonlinguistic basis, has to include something like the restriction or the relativization given above. There is no reason to suppose that this restriction or relativization will not apply across the board to principles of nonlinguistic as well as linguistic-based ascriptions. But if the disease can be cured without a specific reference to language, then it is not likely to have been a specifically linguistic disease to begin with.

Kripke would disagree with the entire line of argument I have been developing, for he would hold that there is a *crucial* difference between Pierre's situation and Martin's that would separate them in a fundamental way. At one point he remarks:

> We have formulated the disquotational principles so they need not lead us to attribute belief as long as we have grounds to suspect conceptual or linguistic confusion . . . (p. 276, n. 23).

Pierre, according to Kripke, is not involved in any conceptual or linguistic confusion. Martin, of course, is not involved in any linguistic confusion either, if by a linguistic confusion one means, as Kripke seems to mean, a confusion that can be resolved by instruction in a single language. Is Martin conceptually confused? It seems not, for his command of concepts is unimpaired, even when he is drunk. So the cases remain unseparated.

There is, however, a way in which the two cases are different, and it is precisely this difference that Kripke must exploit to keep his puzzle afloat. This difference is alluded to in the following passage:

> A straightforward application of the principles of translation and disquotation to all Pierre's utterances, French and English, yields the result that Pierre holds inconsistent beliefs, that logic alone should teach him that one of his beliefs is false. (266–7)

Logic alone will not help Martin either, for, recall, in his drunken states he never reflects upon his sober beliefs, and when sober never thinks about his drunken beliefs. Later, in discussing a variant on the Pierre example, Kripke expands on what he has in mind by saying:

> Yet no amount of pure logic or semantic introspection suffices for him to discover his error. (268–9)

But no amount of *semantic* introspection will help Martin either.

Perhaps we can bring out what Kripke has in mind by supposing that both Pierre and Martin experience what I will call doxastic/logical illumination, that is, each becomes fully aware of his belief system and fully aware of the logical relations between the total set of beliefs that make it up. We will also assume that both Pierre and Martin are logically and semantically articulate; that is, they command the logical and semantic notions needed to describe their doxastic situation. Under such illumination, Martin would become aware that his beliefs are divided; Pierre need not become aware of this. This certainly is a difference, and it suggests that Pierre's divided beliefs are more deeply entrenched than Martin's and therefore that the puzzle concerning his beliefs is deeper than the puzzle concerning Martin's.

But there is confusion here. In a certain way Pierre's divided beliefs *are* more deeply entrenched than Martin's.[8] Pierre, we might say, is in possession of a *coherent world picture* – the world could have been exactly as he conceives it to be, namely, the world could have contained two cities, one named "Londres," the other "London," the one pretty, the other not, etc. All the same, neither Pierre nor Martin has a *coherent picture of the world.* As we have seen, the supposed puzzle about belief arises because a person can have divided beliefs about the world, and is solved by putting a natural restriction on the disquotation principle in recognition of this. The same natural restriction serves for both Pierre and Martin, and for the same reason. There is no deeper puzzle left over. I'll return to this matter in the final section of this essay.

One last point before moving on to Ruth. I think that, to some extent, the puzzle about belief has remained alive because it has been taken to be a puzzle for reference theory. Thus, when various reference theorists attempt to solve it, all the old familiar objections and counterobjections, replies and counterreplies reemerge. My opinion, as indicated above, is that the puzzle about belief is only incidentally a puzzle for a reference theory. A particular theory of reference will be embarrassed by this puzzle only if it somehow precludes the possibility of divided beliefs concerning things referred to by proper names. A sense theory will not be embarrassed in this way, since it allows that expressions with distinct senses may – unknown to the users of these expressions – have the same reference. A pure reference theory will not be embarrassed, since it allows – again unknown to users of these expressions – that two names with the same reference may be connected to that referent by different causal and historical chains. Whatever difficulties Kripke's pure reference theory might have (and they seem to be considerable), he is wrong in thinking that this is one of them.

Ruth

Essays in a Festschrift are intended to honor the person to whom it is dedicated. They are presented as gifts. In a series of articles,[9] Ruth Marcus has offered her own solution to Kripke's puzzle about belief, and if what I have said above is correct, I have deprived her of her subject matter. Some gift! In fact, however, treating her essays as simply (or even primarily) about Kripke's puzzle misses their fundamental importance and originality. The central theme of her essays is to challenge the idea that belief ascription is, in the primary standard case, the ascription of a belief *de dicto*. In her words, "There are failings in such language-centered . . . accounts of belief" (137).

We have already noted that Kripke restricts his discussion to beliefs *de dicto*. We did not note his rather dismissive attitude toward ascriptions of belief *de re*. Twice when he speaks of a *de re* reading of a belief ascription, he adds phrases questioning their sense ["if it makes sense at all" and "if they make sense" (242)]. For Kripke, a belief – or, at least, the kind of belief he is discussing relative to Pierre – is a relationship between a believer and a *dictum*. How that relationship is spelled out depends on how the notion of a dictum is itself understood. It could be a sentence (token or type), an assertion (a datable speech act), or a proposition (itself a notion that can be understood in various ways). But a dictum, as Kripke understands the notion, is *linguistic* or at least *quasi-linguistic* in character. It is something to which the translation principle can apply. Marcus rejects this account of belief. She denies, that is, that having a belief *de dicto* is the essential or primary mode of having a belief. More strongly, she holds – as she must if her position has application to Kripke's puzzle – that the *de dicto* interpretation of beliefs gives the wrong reading of Pierre's beliefs. For Marcus, having a belief is fundamentally different from standing in a relationship to a linguistic entity – unless, that is, the belief is about a linguistic entity.[10]

Marcus offers four criticisms of what she calls the "language-centered concept of belief." First, as she correctly notes, we sometimes attribute beliefs to animals without attributing a corresponding linguistic capacity to them. I do not find this move decisive. It does not seem far fetched, ad hoc, etc., to treat the ascription of beliefs to animals as simply as analogical extension of our normal practice of belief ascription. Much the same response is possible for Marcus's second claim, namely, that the language-centered concept of belief is not applicable to unconscious beliefs. First, maybe it is – at least sometimes – and when it isn't, we can again treat such ascriptions as analogical extensions of standard ascriptions.

Marcus's third criticism carries more weight. There are bases for ascribing beliefs to people other than noting their assent to dicta. Sometimes we ascribe beliefs on the basis of behavior; indeed these ascriptions sometimes override ascriptions based upon observed sincere assent. As Marcus puts it, "Actions may belie our most sincerely reported beliefs" (138). We may also note, though Marcus might resist this claim, that we can ascribe divided beliefs to Martin solely on the basis of his actions. Finally, related to her third criticism, Marcus has a fourth: Sometimes we examine our own actions to discover what we believe. But, as she tells us, "Verbalization as a necessary *condition* of believing precludes our discovering and then reporting *what* we believe" (139).

In place of the language-centered concept of belief, she suggests that "a proper object of believing is a possible state of affairs" (1983, 330). Dividing this in half: The claim that believing is a relationship between a believer and a *state of affairs* provides her alternative to the language-centered concept of belief she rejects. Her more restrictive claim that this must be a *possible* state of affairs is needed to establish her stronger thesis that supposed beliefs in the impossible are no beliefs at all. Clearly, as the title "Rationality and Believing the Impossible" shows, it is the second, more radical, proposal that is dearest to her heart. In what follows, I will argue that her defense of a world-centered over a word-centered conception of belief is all that is needed for the solution of Kripke's puzzle. Resting her criticism on the stronger thesis that contradictory beliefs are not possible is like training a canon on a fly.

Marcus's "world-centered" or "object-centered" account of believing is first presented informally:

> Believing is understood to be a relation between a subject or agent and a state of affairs (not necessarily actual) but which has actual objects as constituents. We may think of states of affairs as ordered structures of actual objects which include individuals as well as properties and relations. (139)

Her more complete formulation combines this world-centered or object-centered approach with a dispositional analysis of believing.

D *x* believes that *S* just in case under certain *agent-centered circumstances* including *x*'s desires and needs as well as *external circumstances*, *x* is disposed to act as if S, the actual or non-actual states of affairs, obtains. (140)

Notice that in neither statement does she mention that the state of affairs must be a possible state of affairs. That is an independent matter, and comes later.

It must be said that Marcus does very little by way of explaining the key ideas in this account of believing. What she does instead is illustrate the force of her position by exhibiting its advantages over competing language-centered accounts of belief. She first argues that it is not subject to the set of criticisms that she leveled against the language-centered account of belief – that is, it allows for straightforward ascriptions of beliefs to animals; it allows for ascriptions of unconscious beliefs; it allows us to refuse to ascribe a belief to a sincere linguistically competent assenter when that person's actions belie his assent, and so forth. Second, she argues that her world-centered theory provides solutions to the standard set of Fregean problems whose solution serves as a success condition on any theory of reference.

There is not space to consider the merits of this position in detail; instead, I wish to make a very general point about the relationship between Marcus's account of belief and Kripke's puzzle. Toward the end of the previous section I insisted on a difference between saying that Pierre has a consistent world picture (which is true) and that Pierre has a consistent picture of the world (which is false). I needed this contrast in order to hold that Pierre is a divided believer even though he did have a consistent world picture. If, as Marcus suggests, *de dicto* ascriptions of belief can only be made under broader *de re*-like constraints,[11] then it is no longer difficult to see how a person can have a consistent world picture – one that survives doxastic illumination – yet still be a divided believer. If the argument of the previous section is correct, this is all that is needed for the diagnosis of Kripke's puzzle, and all that is needed for specifying natural constraints on the disquotational principle to avoid it.

Setting aside questions of detail – and also setting aside Marcus's more radical claim that it is not possible to believe the impossible – I think that she has done two things: First, she has provided a persuasive case that ascription of beliefs *de dicto* can only be made under *de re* constraints; second, she has given a plausible account of the form that such *de re* constraints should take. This is enough to solve Kripke's puzzle about belief. Of course, from Marcus's point of view, this may seem disappointing, because the puzzle gets solved before her story gets interesting. Maybe so, but her views about believing the impossible strike me as deeper than the Kripke puzzle, and would benefit from being developed independently of it.

NOTES

1. Saul Kripke, "A Puzzle about Belief," in *Meaning and Use*, A. Margalit (ed.); (Dordrecht: D. Reidel, 1979), pp. 239–83.
2. Kripke's hypothetical puzzle involves a single person unknowingly having two names for the same city. In the following example, millions of people knowingly have the same name for two different cities. The following is true:

   Italians call Munich and Monaco, Monaco.

   That translates into Italian as:

   Gli italiani chiamono Monaco e Monaco, Monaco.

   This is odd, but still truth-preserving. But the following is also true of Italians:

   Italians believe that Munich is not Monaco.

   That translates:

   Gli italiani credono che Monaco non sia Monaco,

   which is false in attributing a contradictory belief to millions of geographically knowledgeable Italians.
3. Saul A. Kripke, *Naming and Necessity* (Cambridge, Mass.: Harvard University Press, 1980).
4. Immanuel Kant, *Prolegomena to Any Future Metaphysics*, trans. Lewis White Beck (New York: The Liberal Arts Press, 1950), p. 25.
5. An apparent lack of parallelism is discussed in detail later in this paper.
6. Here I am assuming that the ascription of belief to someone does not entail that the person is currently entertaining or reflecting upon this belief. Beliefs, I am assuming, are in some broad sense dispositional. Martin's disposition to speak and act in different ways under different circumstances exhibits the division in his belief system.
7. This is a mere suggestion. Nothing in this paper depends upon the correctness of this relativized version of the disquotational principle. The restricted version of the disquotational principle offered in the text does all the work that needs to be done.
8. Although there is a difference between the level of entrenchment of Pierre's divided beliefs and Martin's, the difference is less than it may first appear. Under doxastic illumination, Pierre would not see that his beliefs *are* divided, but he could come to see that they *might* be. Depending upon the character of his other beliefs, he might come to see that he has no good reason for ruling out the possibility that 'London' and 'Londres' are names for the same place, and thus no good reason for ruling out the possibility that his beliefs are divided. As a rational agent, this should lead Pierre to suspend both beliefs until the matter is settled. Under doxastic illumination, Martin winds up in the same situation, for, as the story is told, nothing guarantees that his belief system, when illuminated, will favor one of his competing beliefs about his sexual prowess over the other.
9. Ruth Barcan Marcus, "A Proposed Solution to a Puzzle about Belief," *Midwest Studies in Philosophy*, ed. P. French et al. 6 (1981): 501–10; "Rationality and

Believing the Impossible,'' *Journal of Philosophy* 81 (1983): 321–39; ''Some Revisionary Proposals about Belief and Believing,'' *Philosophy and Phenomenological Research* 50 (1990): 133–153. Unless otherwise indicated, all references are to the third of these articles.

10. For example, ''Pierre believes that French words sound prettier than English words.''
11. Here I use the awkward expression ''*de re*–like constraints'' to avoid attributing to Marcus a simple *de re* (wide-scope) interpretation of Kripke's target sentences, 'Pierre believes that London is pretty', and 'Pierre believes that London is not pretty.' In rejecting Kripke's *de dicto* analysis, Marcus is not thereby committed to the (implausible) *de re* analysis that Kripke rejects. She is offering an alternative to both. On these matters David Lewis makes the following shrewd observation.

    Kripke may stipulate that [his target sentences] are to be interpreted according to a certain analysis, or he may invite us to rely on our intuitions about their truth. But he cannot do both without begging the question in favour of the preferred analysis. However, it is not at all clear to me what Kripke is stipulating. The notion that ordinary language belief sentences divide into *de dicto* and *de re* without residue or overlap, may be part of an oversimplified semantic analysis. (288) David Lewis, ''What Puzzling Pierre Does Not Believe,'' *American Philosophical Quarterly* 59, no. 3 (1981): 283–9.

13

Closure and Consistency

ISAAC LEVI

Two commonly endorsed conditions on the full beliefs (that is to say, the certainties) endorsed by a rational agent at a given time are that they be consistent and that they be closed under logical consequence. Ruth Marcus (1983) has followed H. E. Kyburg (1970) in maintaining that the closure condition should be abandoned and the consistency condition retained. Unlike Kyburg, however, Marcus does not rest her case on concerns about inductive inference and probabilistic reasoning. She appeals, instead, to Saul Kripke's well-known puzzle about belief (Kripke 1979). I can think of no better way of expressing my high respect for her contribution to the contemporary philosophical debate than to explain why I dissent from her view on this score as I have from Kyburg's on previous occasions.

Students of rational belief change acknowledge that systematic accounts of belief change presuppose some view of the structure of the domain of potential states of full belief available to the inquiring agent (Gärdenfors 1988, pp. 139–40; Levi 1991 §§2.2, 2.4). Just as statistical mechanicians presuppose that mechanical systems are in states belonging to some act of potential mechanical states (the phase space) exhibiting a certain kind of structure (of a $6n$-dimensional Euclidian space where n is the number of particles), so too the domain of potential states of belief is presupposed to have a certain kind of structure.

The precise character of this structure is a matter of some controversy. I have argued elsewhere that the domain of potential states of full belief should constitute at the very minimum a Boolean algebra (Levi 1991, §2.2). A Boolean algebra is a set of objects (in this case potential states of full belief) partially ordered by a relation (which I shall call a *consequence relation*) in a way that yields a distributive lattice with a unique minimum element **0**, a unique maximum **1**, and for every element **K** another element ~**K** that is its complement. The potential state **0** has as a consequence every potential state. The potential state **1** is a consequence of every potential state.[1]

Dedicated with professional respect and personal affection to Ruth Marcus. Thanks are due to Akeel Bilgrami and Walter Sinnott-Armstrong for helpful improvements.

Four kinds of changes in states of full belief can be identified:

(I) *Expansions* are shifts from initial $\mathbf{K}_1$ to $\mathbf{K}_2$ where $\mathbf{K}_1$ is a consequence of and, hence, as weak as (and no more informative than) $\mathbf{K}_2$. There is always a potential state **A** such that $\mathbf{K}_2$ is identical with the meet $\mathbf{K}_1 \wedge \mathbf{A}$ of $\mathbf{K}_1$ and **A**. We say that $\mathbf{K}_2$ is the result of expanding $\mathbf{K}_1$ by adding **A**. $\mathbf{K}_2 = (\mathbf{K}_1)^+_\mathbf{A} = \mathbf{K}_1 \wedge \mathbf{A}$.

(II) *Contractions* are shifts from $\mathbf{K}_1$ to $\mathbf{K}_3$, which is weaker than $\mathbf{K}_1$. There is always a **B** such that $\mathbf{K}_1 = (\mathbf{K}_3)^+_\mathbf{B}$. Hence, $\mathbf{K}_3$ is a contraction of $\mathbf{K}_1$ removing **B.** However, there are many other contractions of $\mathbf{K}_1$ removing **B.** It is typically assumed that there is a best such contraction $(\mathbf{K}_1)^-_\mathbf{B}$.

(III) A *replacement* is a shift from $\mathbf{K}_1$ to $\mathbf{K}_4$ whose meet with $\mathbf{K}_1$ is identical with **0**. $\mathbf{K}_4$ can always be represented as the result of contracting by removing ~**C** from $\mathbf{K}_1$ and expanding by adding **C** to the contraction. If the contraction is an optimal one removing ~**C**, so is the replacement.

(IV) A *residual shift* is a shift from $\mathbf{K}_1$ to $\mathbf{K}_5$ which is neither an expansion, a contraction, nor a replacement. A residual shift can always be viewed as an expansion of $\mathbf{K}_1$ adding **A** consistent with $\mathbf{K}_1$ and then contracting by removing **B.**

Stalnaker, Harper, and Gärdenfors have made another type of belief change a popular object of study. Within the framework of the classification just offered, it is a hybrid.

(*) A *revision* of $\mathbf{K}_1$ by adding **D** (distinct from **0**) is a belief state $\mathbf{K}_6$ distinct from **0** that is identical with $(\mathbf{K}_1)^+_\mathbf{D}$ if **D** is consistent with $\mathbf{K}_1$ and is a replacement of ~**D** by **D** otherwise. If the replacements are optimal or **D** is consistent with $\mathbf{K}_1$, the revision is "minimal" and is symbolized as $\mathbf{K}^+_\mathbf{D}$.

I shall suppose that potential states of full belief are truth valued – that is are either error free (= true) or erroneous (= false). If a potential state is error free, so are all its consequences. If a potential state is erroneous, so are all states which are stronger than it (i.e., have it as a consequence). Consequently **0** is erroneous and **1** error free.

An inquirer's state of full belief **K** serves as the inquirer's standard for serious possibility. Potential states whose meets with **K** do not have **0** as a consequence (and, hence, are "consistent" with **K**) are possibly true in the sense that they are seriously possible according to **K**. Those potential states whose meets with **K** have **0** as a consequence are not possibly true in the sense of serious possibility. A deliberating agent should not assign positive credal probability to a potential state being true unless it is seriously possible that it is true. Only potential states of full belief which count as serious possibilities for the agent may be compared with respect to whether it would be better given the agent's goals and values for the one or the other to be true (i.e., error free).[2] These kinds of appraisals are important in practical deliberation and scientific inquiry. They always presuppose a distinction between what is ruled out as not seriously possible and seriously possible conjectures. The inquirer's state of full belief is the standard for drawing this distinction.

Given an inquirer's state of full belief **K**, the inquirer judges all states that are consequences of **K** to be error free and judges all states **K**′ whose meets with **K** have **0** as a consequence to be erroneous. Relative to **K**, the inquirer is in suspense concerning the truth value of every other potential state. The relevance of such judgments is that when contemplating a shift from **K**, the inquirer should be concerned to avoid error. From the inquirer's point of view, replacements import error for sure. Contractions avoid error for sure. Consistent expansions and residual shifts incur a risk of error but not a guarantee of importing error.

Inquirers are also curious. They seek more information rather than less. That is to say, they would prefer ceteris paribus to move to stronger states if this incurred no risk of error. Expansion is the only kind of change that yields a stronger and, hence, more informative state. It incurs risks of error. To avoid this, inquirers should either remain with the status quo or move to a weaker state. So there is need to establish some trade-off between the desire to minimize risk of error and the concern to obtain new information.

I have argued elsewhere (Levi 1991) that even though shifting into the state **0** imports error for certain (as judged relative to the inquirer's initial point of view), sometimes an inquirer inadvertently but legitimately stumbles into such states. When one conducts experiments that yield results in conflict with background information or where the testimony of reliable witnesses contradicts settled assumptions, such inadvertent stumbling into inconsistency takes place. There is an urgent need then to flee from such inconsistency to a consistent state.

Peter Gärdenfors has defended the view that "inconsistency is epistemic hell" (1988, p. 51) so that whenever one imports information inconsistent with initial information into one's state of full belief, one should shift to another consistent state via a minimal revision rather than expanding into inconsistency. This attitude carries on a tradition initiated by R. Stalnaker (1968).

Sometimes the importation of information inconsistent with the initial belief state eventuates in a minimal revision as Gärdenfors says. However, I do not see how this avoids the epistemic hell of inconsistency. Whenever minimal revision is warranted in a case of a conflict between new information and the initial belief state, the process should be seen as involving two steps: expanding into the inconsistent state **0** and then contracting from it to a consistent state retaining the new information and abandoning old information in conflict with it. Gärdenfors (and Stalnaker and Harper before him) appear to prefer thinking of minimal revision differently. There is no need to think of minimal revision as a two-step process. It can be seen as a single transition from one consistent state to another.

If this approach were acceptable, whenever imported information is inconsistent with the initial state, the transformed state should retain the new information and remove some old information. If there are alternative moves that may be legitimate, there is a need to deliberate as to which alternative is best. The question then arises: best for what purpose? The answer is: extricating the inquirer from the inconsistency between the new information and the initial state. Once this is conceded, it has been conceded that the challenge is how to contract from **0**. If minimal revision is chosen, it should then be regarded as a two-step procedure.

There are alternative options. Suppose, for example, I am strolling on the Columbia campus and see Hugh Mellor walking through the gate at 116th and Broadway. Initially I was convinced that Hugh was in Cambridge (having received an e-mail communication from him that very day). If I had engaged in minimal revision, I would have found myself convinced that Hugh Mellor is at Columbia and the e-mail message was some sort of mistake. But there are two other responses imaginable. I could judge that I was suffering from a mirage and remain convinced that Mellor is at Cambridge. Far more likely, however, I would look more closely. By doing that, I would reveal that I was in doubt as to whether Mellor is at Cambridge as my initial convictions implied or is at Columbia as my first sense impression indicated. That is to say, I would have reacted to the conflict between the new information and the old by contracting the old doctrine so as to be in doubt as to whether the new information is true or information conflicting with it is. Relative to that perspective, I could inquire by looking again without begging the question in favor of one alternative or the other. By way of contrast, looking again makes no sense if I have engaged in minimal revision and become convinced that the new information is true or, instead, have remained in the status quo and remained convinced that the new information is false. It is pointless to check the truth of claims that are already settled.

I contend that in this predicament and others like it where new information conflicts with old, we do, indeed, have available to us alternatives to minimal revision. Consequently, a question arises as to how to choose between these alternatives in cases of such conflict. As I have just argued, this question amounts to asking how to contract from **0**?

Two objections may be leveled against this conclusion: (1) Because **0** is an incoherent standard for serious possibility, in deliberation, I could contract from it in countless ways which would lead to the agent's belief state being represented by a maximally consistent set. But no such move is ever countenanced. (2) Since the agent's belief state is inconsistent, the agent could not evaluate the options for contraction from **0** coherently.

The second point is correct. To be in the inconsistent belief state **0** is to be in a state that is useless for distinguishing serious possibility from impossibility and, hence, for deliberation and inquiry. Hence, the inquirer has to deploy a surrogate belief state for deliberation and inquiry. In this special case, we need to countenance a kind of double standard for belief which, I contend, should not be countenanced elsewhere. The inquirer is in an inconsistent state of belief from which he or she seeks release. But for the purpose of deliberating as to what to do, a surrogate is needed. This surrogate or stopgap belief state should, however, be as harmless as it can be. That is to say, it should not beg questions in favor of one potential contraction strategy rather than another and it should allow for the coherent evaluation of the alternative contraction strategies. Otherwise, it should be as weak as is coherently possible.

Observe that we cannot say that the surrogate belief state is the agent's proper belief state, because then there would be no incentive to extricate the inquirer from inconsistency. In the case of conflict, we cannot, I think, avoid conceding that the inquirer is in an inconsistent belief state in the sense that he or she has such an incentive to retreat from inconsistency. To be sure, the inconsistent belief state cannot serve as the standard for serious possibility. Perhaps, then, we should represent the doxastic condition by the pair consisting of **0** and the surrogate belief state. That would seem acceptable. But it is no escape from epistemic hell. It does not avoid believing the impossible.

Returning to the first objection, I suggest that the available contraction strategies are determined by the initial belief state, which was inadvertently expanded into inconsistency by adding information conflicting with old information. Space does not permit elaboration of details; but roughly speaking, contraction strategies can be of three kinds:

(A) Contractions to the initial doctrine or contractions of it removing assumptions concerning the reliability of the processes that yielded the new information. These are illustrated by the case where I regard my vision of Mellor as hallucinatory.

(B) Contractions to a belief state that is formally representable as the expansion of a contraction of the initial doctrine where the contraction removes information conflicting with the new information and the subsequent expansion adds the new information. These correspond to revisions.

(C) Contractions to a belief state that is the join of belief states of type A and type B. These are contractions of the initial belief state removing both the new information and old information conflicting with it. Such responses are illustrated by my questioning that Mellor is in Cambridge and that he is at Columbia and looking again.

None of these contraction strategies yields maximally consistent belief states. That is because the range of options is tied to the process of expansion into inconsistency, which created the trouble in the first place.

How should one evaluate contraction strategies of these kinds. We cannot compare them with respect to risk of error. Since we have an inadequate standard for serious possibility, we cannot evaluate risk of error either. But we can assess the informational values (i.e., the value of the information carried by the alternative contractions for the purpose of explanation, prediction, and practical deliberation) and seek a contraction that maximizes some assessment of such values. (The details of this approach are discussed in Levi 1991, §4.8).

Gärdenfors urges us to avoid **0** but at least acknowledges it to be a potential state of full belief. **0** is a state to which inquirers are conceptually capable of shifting. Gärdenfors claims that rational agents should not do so but not that it is not an option. Ruth Marcus is more radical. (All references here will be to Marcus 1983). She does not want to countenance **0** as a potential state of full belief at all. But the argument just advanced against mandating minimal revision when something inconsistent with the initial doctrine is added to the evidence argues against Marcus's view as well. If we do not allow for the conceptual ability to shift to **0**, how are we to give a systematic account of changes of states of full belief which ensue when information is added that conflicts with the initial state?

I have argued for allowing rational agents to believe the impossible by appealing to cases where inquirers (inadvertently) add new information to their evolving doctrines conflicting with it. Marcus does not consider this problem. I shall not speculate how she would react to it. The interesting question is whether Marcus is right in prohibiting believing the impossible in situations where new information is not in conflict with old. I think she is although my reasons are different from hers. We all know an agent's linguistic behavior may display a formal inconsistency just as an agent's "assents" may fail to be closed under logical consequence. Marcus maintains that an agent cannot believe an impossibility although the agent can linquistically assent to one because of some sort of confusion or emotional disturbance. She insists, however, that even a rational angel free of confusion or psychic storms may fail to believe the consequences of what he or she believes. In my view, rational agents ought always to believe the consequences of what they believe. They can, under certain circumstances, believe the impossible, as I have already argued. But, as already indicated, I agree with Marcus that when the cases of importing information conflicting with initial information are ignored, inconsistency in belief should never arise. Nonetheless, I insist that an inquirer's full beliefs ought to be closed under consequence. To join the issue with Marcus on the question of closure, some further preliminaries should be reviewed.

If *X* believes that Albany is north of New York, *X*'s belief state is one that has the potential belief state represented by the sentence "Albany is north of New York" as a consequence. In saying this, I did use a sentential representation of potential states of full belief; but this sentential representation did not presuppose any sentencelike object of belief. In particular, I did not presuppose that *X* has a disposition to assent affirmatively to "Is Albany north of New York?" or to any other interrogative. *X* may fail to use any sort of language at all. I did presuppose that we have the ability to represent some of *X*'s potential states of belief linguistically just as mechanicians use points in phase space to represent mechanical states. But the use of language by linguists, psychologists, social scientists and others to represent the states of full belief *X* can shift to is quite different from *X*'s ability or lack of ability to display his current state linguistically. For the most part, I take for granted that *X* has the ability to reveal his belief states through linguistic as well as other behavior. But Marcus is surely right in maintaining that even if *X* lacked the ability to manifest his or her beliefs linguistically, *X* might, in principle, be in belief states.

Not all potential states of full belief are representable by sentences in a suitably regimented language **L**. We may, however, assume that sentences in such a language may be mapped into the set of potential states in a given Boolean algebra so that the logical consequence relation of **L** is preserved by the consequence relation for the Boolean algebra and all logical truths of **L** are correlated with the state **1**. Consider then the set of sentences in **L** which represent consequences of potential state **K**. That set will be closed under deduction in **L** and will represent **K** even if no sentence represents **K**. Thus, we can represent potential states by sets of sentences in **L** closed under deduction – i.e., by theories or potential corpora. But we do not need to invoke an appeal to sets of possible worlds. A Boolean algebra of sets of possible worlds isomorphic to the algebra of potential states of full belief would be gratuitous. Potential states of full belief do all the work that sets of possible worlds may legitimately be asked to do.

Even though Marcus is right in maintaining that sentence-like entities are not needed as objects of belief, she continues to think that objects of belief are required. She takes them to be sets of possible worlds or "states of affairs." Here we part company.

What are states of full belief? To obtain some additional clarity about this matter, imagine that Mario's state of full belief changes in the following ways:

(1) At time t_1, Mario believes that Albany is north of New York and that the relation of being north of is transitive.
(2) At time t_2, Mario comes to believe that Montreal is north of Albany.
(3) At time t_3, Mario comes to believe that Montreal is north of New York.

The doxastic changes at t_2 and t_3 are genuine enough. But they are, I submit, strikingly different. At t_2, Mario changed his doxastic commitment as the outcome of inquiry. Being informed by reliable sources, Mario shifted to a state of full belief having the potential state that Montreal is north of Albany as a consequence from a potential state that did not. He acquired genuinely new information. At t_3, no such change took place. That Montreal is north of New York is already a consequence of Mario's state of full belief. In that sense, he already believed it at t_2. But there is another sense in which he had not as yet believed it. Mario had not previously fully recognized that Montreal is north of New York.

Some changes of belief are changes in doxastic commitment. Others are changes in doxastic performance which are not changes in doxastic commitment. Changes in doxastic commitment are what I am regarding as changes in state of full belief. To be in a state of full belief is to have undertaken certain kinds of doxastic commitments–that is to have obligated oneself to acquire the dispositions and abilities that fulfill the commitments undertaken when this is feasible or to seek ways and means to enhance one's capacity to fulfill the commitments at reduced cost. To have acquired the dispositions and abilities fulfilling the commitments is to have changed one's doxastic performance.

Doxastic commitments are like contracts or promises. The agent does something or shifts into a condition that generates an obligation. The analogy is not, however, perfect. Doxastic commitments resemble more closely religious commitments or undertakings. When someone promises to do something knowing that he or she cannot fulfill the promise, the promise is fraudulent. When an inquiring agent *X* comes to fully believe that *H*, *X* undertakes to recognize as true all consequences of the belief state he has shifted to. No one, however, is capable of doing so perfectly. But those who seek to be righteous or to submit their wills to God rarely think they are capable of doing so perfectly. What they are prepared to do is (a) fulfill the commitment insofar as they are currently able and (b) to seek additional ways and means to enhance their capacities. Doxastic commitments resemble such religious undertakings in these respects.

A doxastic performance is an attempt to fulfill a doxastic commitment. If the doxastic commitment is full belief that *H*, a successful performance is a full recognition that *H*. If it is unsuccessful, it is a (failed) attempt at full recognition.

Doxastic performances are behavioral or linguistic dispositions and, perhaps, various neurophysiological conditions. We should distinguish not only between doxastic commitments and doxastic performances but between the latter and their overt (linguistic or behavioral) manifestations. Mario's dispositions to utter or inscribe linguistic expressions and his dispositions act nonlinguistically in response to various provocations are one thing. His utterances, inscriptions, and behavioral responses are another.

Doxastic commitments, performances, and manifestations have all been called "believings." I am restricting 'belief' to the doxastic commitments, calling performances 'attempted recognitions' that are manifested in linguistic behavior (such as assents) as well as nonlinguistic behavior. When an attempted recognition succeeds in fulfilling a doxastic commitment, it is a recognition. I hold no brief for my terminological stipulations. The distinctions are what matter.

Given these distinctions, a rational agent, whether a rational angel who perfectly fulfills his or her commitments or not, fully believes all consequences of his or her current belief state. If the agent is not angelic, however, he or she will fail to recognize all consequences of his or her current belief state. That is to say, the agent's linguistic and nonlinguistic behavior will fail to exhibit obedience to closure requirements; but the agent's beliefs will satisfy closure.

Consequently, when the agent's linguistic and nonlinguistic behavior suggests belief in the impossible, it can be read in one of two ways: (i) The agent is doxastically committed to recognizing **0** as true. That is to say, the agent is in the state of full belief **0**. (ii) The agent's linguistic and nonlinguistic behavior is somehow incoherent but the agent is not in belief state **0**.

Because of emotional instability, self-deception, lack of computational capacity, and the like, an agent may display linguistic or nonlinguistic behavior that normally would be interpreted as recognizing that *S* and also display behavior normally interpreted as recognizing that ~*S*. If these normal construals are endorsed as manifesting fulfillments of the agent's doxastic commitments, the agent is being construed as in the belief state **0**. The inquirer got there because of confusion. Hence, his or her being there is in that sense inadvertent but not the sort of inadvertence that obtains when one expands into inconsistency through relying on the testimony of the senses or expert witnesses. Rational angels who perfectly fulfill doxastic commitments can inadvertently stumble into inconsistency in the latter way but not the former.

It is undeniable that agents do exhibit the sort of behavior sketched in the previous paragraph and that we can sometimes explain it as due to emotional stress, lack of computational capacity, and other such disabilities. But it would be wrong to suggest that appeal to such confusion either explains or excuses doxastic commitment to **0**. For this reason, it seems doubtful that we should construe him or her as believing **0** – that is, as being committed to recognizing that *S* is true, that ~*S* is true and, hence, that *S*&~*S* is true. We might admit that the agent's linguistic and nonlinguistic behavior manifests attempts to fulfill such commitments; but at least some of these performances fail. Any behavior linguistic or otherwise which seems to manifest recognition that *S*&~*S* does not genuinely do so. The agent's performance fails to match his or her commitment due, of course, to the confusion, emotional instability, and lack of computational capacity mentioned before. The entire point of citing failures of computational capacity, emotional instability, and the like is to excuse the agent for failing to fulfill his or her doxastic commitments – that is, for not doing what he or she ought to have done. This point would be undermined by interpreting the agent as fulfilling an inconsistent commitment. Only if the behavioral and linguistic data suggested that an inquiring agent clearheadedly, resolutely, and deliberately attempted to recognize that *S*, recognize that ~*S* and recognize that *S*&~*S* would we be prepared to conclude that all of these attempts were successful and that we should interpret the agent as believing the impossible. Thus, I am inclined to agree with Marcus that reference to confusion, failure of computational capacity, emotional instability, and other impediments to clear thinking does not account for why agents sometimes believe the impossible but only for the incoherence of the linguistic and nonlinguistic behavior that manifests attempted fulfillments of doxastic commitments. This does not mean, however, that the inquirer cannot believe the impossible. When the inquirer inadvertently stumbles into inconsistency via routine expansion consulting the testimony of the senses and expert witnesses, his expansion into inconsistency is not, in general, the result of intellectual confusion or emotional instability.

According to the position I have taken here, if we follow this interpretative path and refuse to credit the agent with believing the impossible, we must also insist that either his apparent recognition that *S* or his apparent recognition that ~*S* was also a failed performance and not a genuine recognition (that is, fulfillment of a doxastic commitment). The reason is that if the inquirer fully believes that *S* and also fully believes that ~*S*, he must be in the state of belief **0** and, hence, believe the impossible. That is the only state which has the potential states that *S* and that ~*S* as consequences.

Marcus, however, thinks that the inquirer *X* may fully believe that *S* and also fully believe that ~*S* without fully believing the impossible state of full belief **0**. She rejects the idea that belief is closed under consequence. The difference in our views is reflected in the ways we analyze the claim that *X* believes that *S*.

I contend that to believe that *S* is to be in a state of full belief which has the potential state that **S** as a consequence. Similarly, to believe that ~*S* is to be in a state of full belief which has **~S** as a consequence. Since the claim is that the current state of full belief is the same in both cases, it must be the case that the agent has the potential state that is the meet of **S** and **~S** as a consequence; for given the Boolean structure of the space of potential states, any state that has both **S** and **~S** as consequences must have the meet as a consequence.

Marcus rejects the construal of the claim that *X* believes that *S* on which this argument is predicated. She thinks that to believe that *S* is to have some sort of doxastic relation to a state of affairs or set of possible worlds. Sets of possible worlds constitute a Boolean algebra with respect to the superset relation: **0** may be correlated with the empty set, **1** with the domain of possible worlds. One potential state is a consequence of the other if and only if the set of worlds containing the alleged consequence contains the set of worlds which is stronger than it. From the point of view of a student of belief change, it is difficult to say what the advantage is to having a Boolean algebra as the doppelganger to the Boolean algebra of potential states of full belief. The construction seems gratuitous.

Marcus, however, finds an advantage. If we adopt the approach I favor, an inquirer must believe all the consequences of his state of full belief. If we supposed that a potential state of full belief is represented by a set of possible worlds, we are not required to do so. Marcus is very anxious to insist that an agent's belief state should be consistent but need not be closed under consequence.

To be sure, a potential state will not be understood in precisely the same way as I have understood it. It cannot serve as the agent's standard for serious possibility. The reason is this. A potential state of full belief is a serious possibility relative to a potential state of belief **K** serving as a standard for serious possibility if and only if the filter it generates (its set of consequences) does not contain the complement **~K** of **K**. A sentence S represents such a serious possibility if and only if it is consistent with the corpus K representing **K**. If one follows Marcus's approach, one might say that the state of affairs characterized by a set of possible worlds is a serious possibility according to a state of belief **K** if and only if the set of possible worlds which is the object of that state of belief has a nonempty intersection with the set of possible worlds constituting the state of affairs associated with the belief state represented by S.

No matter whether one uses the current belief state to distinguish between potential states of belief or between states of affairs with respect to serious possibility, the set of potential states of full belief (states of affairs) that are serious possibilities relative to the standard of serious possibility **K** consists of all and only states whose meets with **K** are not identical with **0** or, equivalently, consists of all and only sets of possible worlds whose intersections with **K** are not empty. It follows that all and only states that are consequences of **K** are such that their complements are not seriously possible.

I would say that such consequences are fully believed relative to **K** and that the inquirer is committed to fully recognizing all of them. Marcus denies this. I conclude from this that for Marcus, what an inquirer fully believes cannot serve as a standard for serious possibility.

I have supposed that an agent's standard of serious possibility is of great relevance to the agent's deliberations in making decisions. When evaluating hypotheses as more or less probable or with respect to whether it is better or worse relative to one's goals that they be true, the hypotheses evaluated in these respects are conjectures that are judged possibly true or possibly false. The sense of possibility relevant here is not truth in some possible world but some notion of possibility akin to subjective or epistemic probability. It is what I call *serious possibility*. Any potential state of full belief (state of affairs) which is not a serious possibility should be assigned **0** subjective probability by a rational agent who is in belief state **K** and should not be assigned a utility relative to **K**.

If this vision of the function of a state of full belief is endorsed, then a person who believes that *S* and, hence, rules out as not seriously possible the potential state of full belief or the associated state of affairs (i.e., set of possible worlds) represented by the sentence ~S and also believes that ~*S* so that the potential state of full belief and associated set of possible worlds represented by S is ruled out as a serious possibility is committed to ruling out the set of all possible worlds and the potential state **1** as a serious possibility. Such an agent is in belief state **0**. The state of affairs associated with that belief state is the empty set of possible worlds. To be in such a predicament is an excellent reason for contracting the belief state. But it is also an excellent reason for concluding that a person who believes that *S* and believes that ~*S* does believe the impossible. If Marcus rejects this last claim, she is denying that a state of full belief is a standard of serious possibility. She should then explain why the notion of full belief she has in mind is of any relevance to practical deliberation or scientific inquiry.[3]

Marcus does offer us a sketch of a view of the importance of the notion of belief which might be entertained as a response to this challenge. We display linguistic and nonlinguistic behaviors that are sometimes rationally consonant with one another. She thinks that attributing beliefs to the agent can explain such consonance when it occurs.

If we took seriously the emphasis Marcus places on explanation, we should have to deny any value to belief attribution. There is always dissonance (both within linguistic behavior and between linguistic and other forms of behavior). Uttering sentence **A** may be consonant with uttering sentence **B**. But such behaviors will quite likely be dissonant with other behavior such as failing to assent to or even dissenting from a sentence which is a logical consequence of **A** and **B** difficult for the agent to recognize. How, in the light of Marcus's views on the differences between beliefs and assents, are we to explain the consonant assents to **A** and **B** as due to beliefs rather than explaining the locally consonant dissent from the consequence as due to beliefs? Unless there is complete freedom of the linguistic and nonlinguistic behaviors from global dissonance, Marcus will have to decide which locally consonant behaviors to explain and which to avoid explaining. So explaining locally consonant behavior cannot be the sole motivation for introducing belief attributions. Otherwise belief attribution is useless for explanation of linguistic and nonlinguistic behaviors as long as there is any dissonance.

It is not as though we can say that beliefs play a role in explanation under idealized circumstances. We are not rational angels. Our behavior is infected with dissonance. But we are sufficiently rational so that invoking a belief–desire model under ideal circumstances may model the behavior of flesh and blood to a good degree of approximation. In the case of ideal explanation in natural science, inquirers do not rest content with the idealizations but seek to improve and correct the idealized theory to bring the ideal closer to the real. Often the result is a substantial transformation of the original theory. If this is the way in which idealization works in ideal rational explanations, we would end up replacing our models of rationality with alternative models not even remotely similar to the initial "folk psychological" view. Yet, it is clear that abandoning belief–desire models for rational choice is not widely endorsed. I conclude that those who do not endorse abandonment should give up the idea that such models serve a primarily explanatory function.

Models of rational belief, desire, and choice are used chiefly for prescription. Beliefs are undertakings and, as such, generate obligations. Models of rationality identify what these obligations are. They are not chiefly used to explain intentional behavior. There is nothing in a behavioral or linguistic disposition or in a neurophysiological state described in physical or biological terms which identifies whether these dispositions attempt to fulfill obligations generated by doxastic undertakings and, if they do, what the undertakings are and whether they succeed. Answers to these questions are themselves implicated in judgments as to the obligations shouldered by the agent and cannot, for this reason, be teased out of information about physics and biology alone. The gap between the intentional and the nonintentional is of a piece with the fact-value dichotomy.

A rational angel could and would completely fulfill his or her or its doxastic commitments. There would be no failure to recognize consequences of his, her, or its belief state because of lack of computational capacity, failure of memory, or emotional storm. Rational angels are logically omniscient. But they need not be omniscient. They might still need to engage in inquiry to improve their doxastic commitments. But they would not need therapy to improve their rational health. They would not need to study logic or mathematics, to exploit computers, tables for chi square, or undergo psychotherapy to remove impediments to clear thinking.

Both human agents and corporate agents (and any other agents we may identify such as, more doubtfully, some animals or automata) are not rational angels. We need not only to inquire but to undergo rational therapy appealing to the several helping disciplines (logic, mathematics, psychotherapy, computer technology, etc.) for help in enhancing our capacity.

To be a rational agent does not require that one be a rational angel but only that one undertake to be one. Consequently, when *X*'s state of full belief **K** is represented linguistically by the logically closed set **K** of sentences in **L**, it is not being supposed that *X* recognizes as true the potential state of full belief represented by sentence **S** for every **S** in **K** but only that *X* is committed to such recognition.

Notice, however, that if *X* is in the state of full belief **K**, *X* judges or believes true every potential state of full belief which is a consequence of **K** even though *X* will not recognize as true all of these consequences. This claim is merely an artifact of my analysis of "*X* believes that *S*" and does not attribute to *X* the capacities of a rational angel. As I suggested earlier, the belief attribution asserts that the potential state **S** is a consequence of *X*'s state of full belief. What is philosophically important about the analysis is that it shows how one can dispense with objects of belief altogether. In spite of her dismissal of linguistic or quasi-linguistic objects of belief, Marcus still thinks of beliefs as having objects such as possible states of affairs or (nonempty) sets of possible worlds. In my view, there is no need for such objects in providing an account of belief change or of the obligations generated by states of full belief, states of probability judgment, value commitments, and choices.

Having said all of this, it remains true that if we could assume that the agents under consideration are rational angels perfectly fulfilling their doxastic obligations, we might explain their behavior by appeal to their beliefs. Marcus seems to think that attributing belief in the impossible cannot have even this attenuated explanatory role. I disagree. A rational angel, as noted, can legitimately import information inconsistent with his initial doctrine into it and then seek ways and means to extricate himself from difficulty. In this setting, attributing belief in the impossible plays an important explanatory role in accounting for the kinds of changes in doxastic commitments the inquirer seeks to make. And it does so without betraying the idea that the inquirer's belief state functions as his or her standard for serious possibility. To the contrary, to the extent that Marcus is prepared to dismiss that function, she is abandoning her own conviction that it is possible to explain – at least in the case of ideally rational angels – linguistic and nonlinguistic behavior in such predicaments in terms of belief and desire. Since that is the only explanatory role beliefs can plausibly play, I am inclined to think that when all its said and done, taking beliefs as having sets of possible worlds as objects does not allow us to deny that the inquirer's belief state is closed under the consequence relation between potential belief states.

Thus far, however, nothing has been said about the considerations that persuaded Marcus to abandon the closure condition on rational belief. To sustain my position, some attention needs to be devoted to Kripke's so-called puzzle about belief.

Suppose Unlucky Pierre utters the sentence 'Londres est jolie' and also the sentence 'London is not pretty'. Marcus is interested in what these linguistic utterances tell us about Pierre's beliefs. Keep in mind that here linguistic behavior is being used as data constituting relevant evidence of Pierre's current state of full belief. We must not confuse the language Pierre uses to manifest his doxastic dispositions (which may or may not be fulfilling his doxastic obligations) with the language we (including, perhaps, Pierre) use to partially describe his belief state. The latter language L may be a regimented fragment of the language Pierre speaks, it may be a regimented fragment of some other language or it may be a more formalized language. What is crucial to appreciate is that the resources of the language used to manifest doxastic and other attitudinal dispositions (such as desires, hopes, etc.) need not and, indeed, should not be incorporated into the regimented language. For example, someone might utter the words 'Quantum mechanics is true' without being in a position to specify a single postulate of quantum mechanics. It is an open question as to whether we would be prepared to attribute to that person a doxastic commitment to the truth of quantum mechanics. But if we did, we should not say that the person is in a state of full belief having as a consequence the potential state that quantum mechanics is true. That is to say, we should not do so unless we take 'quantum mechanics is true' to be an abbreviation of a claim to the effect that systems of elementary particles are quantum mechanical systems and where being a quantum mechanical system is explicated by reference to a specific system of axioms. This is a respect (one of many) in which the regimented language whose sentences are mapped into the potential states of full belief so that logical consequence preserves the consequence relation between potential states must be pruned of resources of vital importance in the linguistic manifestation of doxastic performances but which are inconvenient in the representation of potential states of full belief.

The relevance of these points to our concerns is that when its force is fully appreciated, we must see that the "disquotational principle" advanced by Kripke and its modification advanced by Marcus are untenable. These principles license sufficient conditions for inferring from a speaker's "assenting" to a sentence P in a language the speaker is using the speaker's believing that *P*. But if the sentence P in the speaker's language is one that fails to appear in the regimented variant, the inference must be illegitimate.

Consider the case of Unlucky Pierre once more. He has a disposition to assent to 'Londres est jolie' which, according to Kripke, translates into 'London is pretty'. This licenses the judgment, according to both Marcus and Kripke, that Pierre believes that London is pretty. Similarly, Unlucky Pierre has a disposition to assent to 'London is not pretty' which licenses the conclusion that Pierre believes that London is not pretty.

But both Marcus and Kripke subscribe to the view that the terms 'London' and 'Londres' are just tags directly referring to the same city. To be sure, both of them insist that the result about Pierre could be obtained by treating 'London' and 'Londres' as abbreviated definite descriptions provided that the descriptions were translations of one another. We do not need to worry about the correctness of this claim. Even if 'London' and 'Londres' are given "Millian" semantics of the sort Marcus and Kripke advocate, we should ask what ramifications this has for using 'London is pretty' and 'London is not pretty' to represent potential states of full belief. Raising this question puts us well on the way to a response to Kripke's puzzle alternative to Marcus's.

Suppose that Pierre had begun in the state of ignorance **1** represented by the corpus or theory **1** consisting of sentences which are members of any corpus representing a potential state of full belief. Such a corpus would contain neither 'London is pretty' nor 'London is not pretty'. Suppose further that Pierre "expands" his belief state by shifting to one represented by the result of adding 'London is pretty' to **1** and forming the closure $\mathbf{K}_1$.

Contrast this with the result of shifting to the belief state represented by the result of adding 'something is pretty' to **1** and forming the closure $\mathbf{K}_2$.

Is $\mathbf{K}_1$ the same as or different than $\mathbf{K}_2$? Are the corresponding belief states $\mathbf{K}_1$ and $\mathbf{K}_2$ the same or different?

The appropriate answer to give depends on whether 'London' is taken to be an abbreviation for some definite description such as 'The largest city straddling the Thames' or is not. If it is, then strictly speaking $\mathbf{K}_1$ is the product of adding 'The largest city straddling the Thames is pretty' to **1** and forming the closure, which is clearly different from $\mathbf{K}_2$. The disquotational thesis fails.

But according to the Marcus–Kripke view, 'London' is a mere "tag." Given the interest in belief change, we should say either that 'London is pretty' does not belong to the regimented language and, hence, is not mapped to some potential state of full belief such as $\mathbf{K}_1$ or is to be understood as a flagged variable used in dropping existential quantifiers in predicate logic and represents the same potential state as 'Something is pretty' does – i.e., $\mathbf{K}_2$. In both cases, the disquotational thesis fails. But if we do not adopt one of these two alternatives, we shall have to say that 'London' carries more information than a tag should carry – information such as that some city is pretty or that some city on the Thames is pretty.

Imagine that Unlucky Pierre utters both "London is not pretty" and "Londres est jolie." We suppose that Pierre is sufficiently clearheaded and logically omniscient so that he makes no elementary mistakes. We are supposing that his current belief state is one that has as consequences potential belief states $\mathbf{K}_3$ and $\mathbf{K}_4$ represented by "London is not pretty" and "Londres est jolie" respectively. If we take the view that proper names are tags, $\mathbf{K}_3$ is representable by the corpus $\mathbf{K}_3$ consisting of the logical consequences of "Something is not pretty," which by hypothesis is the appropriate regimentation of "London is not pretty." $\mathbf{K}_4$ may be represented by an appropriate regimentation of the English translation of 'Londres est jolie.' This translation will have the form "——is pretty" where the blank is replaced by a proper name. Whatever the proper name is, the regimented version is "Something is pretty." Clearly, the meet of $\mathbf{K}_3$ and $\mathbf{K}_4$ which must also be a consequence of Pierre's belief state is not identical with **0**. However, the question arises as to whether the fact that Pierre uttered the English and the French sentences entitles us to conclude that he is committed to utter the conjunction of the English sentence and the English translation of the French sentence. If we are prepared to say this (as Marcus and Kripke are), we need to be more specific about the English translation of 'Londres est jolie'. Keep in mind, however, that our interest here is in obtaining more information about Unlucky Pierre's state of full belief. Once we have made a decision about the English translation of 'Londres est jolie', we may suppose we can form the conjunction of that translation with 'London is not pretty' and ask for the regimentation that represents a belief state that is a consequence of Pierre's current belief state. If 'Londres' is translated as 'London', then the conjunction is 'London is both pretty and not pretty,' which regiments to 'Something is both pretty and not pretty.' This sentence clearly represents **0**.

This decision to translate 'Londres' as 'London' is not a consequence of our treating proper names as tags, nor do Marcus and Kripke, who favor the decision, claim that it is. According to them, it is because both names refer to the same thing or have the same "semantic value." Either this talk of reference and semantic value is just another way to interpret Pierre as believing **0** or such talk has no bearing on interpreting Pierre at all.

The other decision one can make is to refuse to translate 'Londres' as 'London' and to translate it instead as 'London*' or with some other suitable epithet. In that case the regimented version of the conjunction of 'London is not pretty but London* is' will be 'Something is not pretty but something is'. There is no impossibility in this.

How should one interpret Pierre? That is to say, how should we represent his beliefs? I do not think it depends upon whether one thinks proper names are abbreviated descriptions or are not. I have carried on this discussion on the assumption that they are not abbreviated descriptions. But this does not drive one to translate 'Londres' as 'London' for the purpose of representing the belief state to be attributed to Pierre on the basis of his linguistic behavior. It is no doubt true that the default translation for 'Londres' is 'London'; but that default translation cannot automatically be used in all contexts and indeed should not, when being strict about things, be used at all. On the antidescriptionist view, the regimented language used to represent states of full belief must treat proper names as flagged variables. Customary translations tend to treat proper names as abbreviated descriptions. Hence, when 'Londres' is translated as 'London', we are claiming that whatever the descriptions are of which 'Londres' is an abbreviation, their translations are customarily abbreviated by 'London'. This kind of information can be useful when seeking to represent someone's belief states provided we can rest confident that the individual's linguistic and other behavior conforms to customary practice. But Pierre is clearly a deviant character even if he is clearheaded and rational. So we as translators would be foolish to impose the default translation of 'Londres' when seeking to infer Pierre's doxastic commitments.

So it appears that I, like Marcus, have provided a way for us to avoid attributing belief in the impossible to Pierre. Unlike Marcus, I do not deny that Pierre believes the consequences of his state of full belief. It is a consequence of Pierre's belief state that there are at least two things (if he has more information, these things are cities) one of which is pretty and one of which is not. Pierre's corpus is deductively closed. He is committed to recognizing as true all consequences of his belief state. And assuming he has the capacities of a rational angel, he is fulfilling his doxastic commitments so that his linguistic and other behavioral dispositions are successful performances fully meeting his doxastic obligations.

Thus, I do not think that Pierre is so unlucky as Kripke suggests. He is not driven to believe the impossible. On the other hand, I do not think that he is as lucky as Marcus thinks. His beliefs are closed under consequence as well as consistent. Pierre is committed to judging true all the consequences of what he fully believes. His corpus is closed under logical consequence and if he is a rational angel, he fulfills the commitment. Not only his dispositions to assent but all his behavioral dispositions can then cohere with these commitments.

I understand that those who embrace theories of direct reference for proper names will not be happy with this conclusion. They should note carefully, however, that I have not moved to the Fregean view instead. Throughout I have avoided supposing that proper names are abbreviations of descriptions or are somehow to be interpreted on some "cluster theoretic" approach. I have suggested that if we are thinking of proper names being used in linguistic representations of full belief of use in studying belief change along Millian lines, it is best to think of them in normal situations as flagged variables. When we do so, we are driven neither to inconsistency nor to violating closure.

There are, however, many common situations where we are unlucky. These arise when in the course of inquiry we deploy routines for obtaining new information which inadvertently and unexpectedly inject inconsistency into our evolving doctrines–as, for example, in my fictional encounter with Mellor. This can happen even to rational angels who make no computational errors. When that happens, we seek to escape from inconsistency as fast as is feasible. Even so, to investigate critically how we should change beliefs in such predicaments, we ought to acknowledge explicitly that such predicaments do arise and that when they do, we do, indeed, believe the impossible.

NOTES

1. The notion I propose to take as primitive here is that of one potential state of full belief being a consequence of another. Boolean algebras can, of course, be characterized beginning with different primitives. When the domain under consideration consists of potential states of full belief, to say that state y is a consequence of state x is to say that x is at least as strong as, carries at least as much information as y and should gratify curiosity at least as much as y. Given a partial ordering with respect to such a consequence relation, the meet $x \wedge y$ of two potential belief states x and y is the weakest (least strong) belief state that is a consequence both of x and of y. The join $x \vee y$ is the strongest potential belief state that is a consequence of both x and y. When the meets and joins exist for any pair x and y, the partial ordering is a lattice. If for every x, y, and z, (i) $x \vee (y \wedge z) = (x \vee y) \wedge (x \vee z)$ and (ii) $x \wedge (y \vee z) = (x \wedge y) \vee (x \wedge z)$, the lattice is distributive. Assuming the existence of **0** and **1**, if for every x, there is an x^* whose meet with x is **0** and whose join is **1**, the distributive lattice is a Boolean algebra. For a discussion of Boolean algebras, see Rasiowa and Sikorski (1963).
2. We can, of course, wish that matters were different than they are and compare propositions we are sure are false with respect to the value of their being true relative to a revision of the current state of belief, which renders these propositions serious possibilities.
3. This challenge is essentially the same as I think H. E. Kyburg (1970) should meet.

REFERENCES

Gärdenfors, P. 1988. *Knowledge in Flux: Modeling the Dynamics of Epistemic States.* Cambridge, Mass.: MIT Press.

Kripke, S. 1979. A Puzzle about Belief. In *Meaning and Use*, A. Margalit (ed.), Dordrecht: Reidel, pp. 224–83.

Kyburg, H.E. 1970. Conjunctivitis. In *Induction, Acceptance, and Rational Belief.* M. Swain (ed.), Dordrecht: Reidel, pp. 55–82. (Reprinted in *Epistemology and Inference* 1983. Minneapolis: U. of Minnesota Press, pp. 232–54.)

Levi, I. 1991. *The Fixation of Belief and Its Undoing.* Cambridge: Cambridge University Press.

Marcus, R. 1983. Rationality and Believing the Impossible. *Journal of Philosophy* 80: 321–38.

Rasiowa, H., and R. Sikorski. 1963. *The Mathematics of Metamathematics.* Warsaw: Pantswove Wydawnictwo Naukove.

Stalnaker, R. 1968. ''A Theory of Conditionals.'' In N. Rescher (ed.), *Studies in Logical Theory.* Oxford: Blackwell, pp. 98–112.

14

De Re Belief, Action Explanations, and the Essential Indexical

ERNEST SOSA

"Latitudinarianism" concerns *de re* belief and its relation to *de dicto* belief:

(LB) S believes x to have property ϕ if and only if there is an individuating concept α satisfied uniquely by x in a context C, such that, in C, S believes *de dicto* an abstract proposition that predicates ϕ with respect to α (which shall be represented here by '[ϕ, α]').[1]

("LB" is short for "Latitudinarianism of Belief.") The proposition that the cat before me is seated predicates being seated with respect to the individuating concept ⟨the cat now before me⟩. An individuating concept may be viewed as what is expressed by a singular term, such as the definite description 'the cat now before me.' We may here think of the relevant "context C" as constituted by a subject S and a time t. When does a cat x satisfy uniquely the individuating concept ⟨the cat now before me⟩ in that context C? When x is uniquely a cat before S at t. Finally, if S believes *de dicto* (the proposition) that the F is G, then not only the property of being G but also the individuating concept ⟨the F⟩ both enter into his conception of how things stand.

The doctrine can be extended to propositional attitudes in general and to intention in particular. Thus:

(LI) S intends to ϕ x iff there is an individuating concept α satisfied uniquely by x in a context C, such that, in C, S intends *de dicto* a proposition that predicates ϕ'ing with respect to his first-person concept of himself (his concept of himself as himself) and α (in that order).

("LI" is short for "Latitudinarianism of Intention.") If S intends *de dicto* a proposition that predicates ϕ'ing with respect to himself (as himself) and ⟨the F⟩, in that order, then not only the property of ϕ'ing but also the individuating concepts ⟨Myself⟩ and ⟨the F⟩, both enter into his conception of how, according to his intention, things are meant to turn out. Thus if I intend *de*

Here I write not at all about Ruth Marcus's work, but very much in admiration of it and in her honor.

dicto (the proposition) that I feed the cat before me, then not only the property of feeding but also the individuating concepts ⟨Myself⟩ and ⟨the cat before me⟩ both enter into my conception of how things are to turn out.

Let us now consider such latitudinarianism and discuss in its terms the following issues:

(A) The problem of how to explain selective action by a subject *S* on objects that, from the perspective of *S*, are *de dicto* indistinguishable.
(B) The "shell game" problem. Suppose a long-standing object is replaced by an indistinguishable surrogate. Do our relevant beliefs and intentions now refer to the original or to the surrogate?
(C) Basic action and its place in explanation of intentional action.
(D) The problem of the essential indexical.

A

In contrast with latitudinarianism, according to a narrower doctrine of *de re* attitudes you do not attain genuine reference to a particular object merely by having in your thought some concept that denotes that object. To think about, to refer in thought to, the tallest spy, it is not enough simply to have the thought that the tallest spy is bound to be a spy, even supposing that there is such a spy (taller than any other).

(N) For genuine reference, some further special relation must bind the thinker with the object of reference, probably some causal psychological relation such as perception or memory.

"N" stands for this narrower doctrine of *de re* attitudes.

Two sorts of cases have been proposed as problems for latitudinarianism, and evidence for *N*. First there is the problem of the agent *S* who believes *falsely*

that the *F* = the *G*,

since though the *F* exists and so does the *G*, they are actually distinct things. In that case, from the perspective of *S*'s beliefs and intentions, it may well be impossible to distinguish the *F* from the *G*. There is *nothing* that *S* believes or intends with respect to 'the *F*' that he does not believe or intend with respect to 'the *G*.' There is no belief or intention that *S* would appropriately express by '. . . the *F* . . .' not matched by a corresponding belief or intention expressible by '. . . the *G* . . .'.[2]

Suppose that the supermarket security officer believes mistakenly that the man streaking down the aisle as he strews sugar is the pillar of the community, Bernard J. Ortcutt. A quick logician, the officer believes or intends nothing that he would express by '. . . the man streaking down the aisle strewing sugar . . .' unless it is matched by a corresponding belief or intention expressible by '. . . the pillar of the community, Bernard J. Ortcutt . . .'. According to latitudinarianism, then, the officer holds no belief or intention about either the streaker or the pillar without holding a matching attitude about the other. How then do we explain that he nabs the streaker and not the pillar?

The moral we are meant to draw is that there must be some further, more direct and causal relation that connects the officer with the streaker before him, some perceptual relation perhaps, one that does *not* connect the officer equally with the pillar. It is this difference that is meant to explain the officer's nabbing the streaker and not the pillar. And doubtless there *are* differences in causal and perceptual relations involved in our example, but we need not infer that it is some such further relations that constitute the officer's *de re* reference to the streaker. For we can explain the difference within the confines of latitudinarianism, as follows.

While it is true for latitudinarianism that the officer has the same *de re* beliefs about the streaker and the pillar, some such beliefs are especially pertinent to who ends up being nabbed by the officer. And there is a salient difference here as we compare these beliefs about the streaker with the corresponding beliefs about the pillar: namely, that the beliefs are in fact true of the streaker while false of the pillar. For example, the officer believes about both the streaker and the pillar that he runs naked just ahead as he strews sugar down the aisle. But this *de re* belief is of course true only about the streaker and not about the pillar. More generally, an agent *S* will act on an intended object of action *O* only if *S* has *true* beliefs that relate *O* appropriately to actions under *S*'s direct control at the time. Thus in our example *S* might believe: "If I shout, right now, from where I stand, at the streaker (i.e., the pillar) that he is under arrest, he will hear me and stop," but this would be true of the streaker while false of the pillar, which explains in latitudinarian terms why the officer arrests the streaker and not the pillar. Such an explanation of the officer's differential success is unaffected by the fact that the streaker and the pillar are indistinguishable from the perspective of the officer's body of beliefs, which also implies the indistinguishability of the two in respect of *de re* beliefs held by the officer. What matters is that some important such beliefs are *true* of the streaker and *false* of the pillar, which determines, for example, who is and who is not within nabbing range.[3]

B

According to a second objection, latitudinarianism is refuted by the "shell game" problem:

> Suppose there was earlier a pepper mill (x) to the left of S, and that S saw it and (mis)took it to be a full salt-shaker. Later the pepper mill was removed and replaced with a full salt-shaker (y), without S's realizing it and without S's ever perceiving or knowing of y. S now reaches to the left in order to pick up the object he believes to be there (the pepper mill, presumably, the one that he erroneously believed and believes to be a full salt-shaker). S does pick up the object that is now there: as it happens, fortuitously, a full salt-shaker, but not the object S was reaching for, the object S believed and believes to be there.

It has been objected that in this example the conditions laid down by latitudinarianism (LI above) for S to intend to ϕ y (the salt-shaker now to S's left) are all satisfied.[4] For, we are told ⟨the salt-shaker to the left⟩ is an individuating concept α such that, in the context of that example (C):

(i) α is uniquely satisfied by the salt-shaker y in that context C; and
(ii) S, in C, intends *de dicto* a proposition that predicates reaching with respect to himself and α in that order.

Yet, according to the objection, the agent S really intends to reach the pepper mill rather than the salt-shaker, since it is the pepper mill that he saw and remembers and believes to be still there (even though he misperceived it and erroneously believed and believes it to be a full salt-shaker).

Here again the example is supposed to cast doubt on any latitudinarian conception of reference in thought, and to promote a narrower conception of the sort indicated by N above. It is supposed that through a combination of memory and perception the agent S is genuinely related in thought to the pepper mill and not to the "imposter" salt-shaker, which merely happens to fulfill by accident the abstract conceptual content of S's intention.

For a solution to the puzzle which preserves the spirit of our latitudinarian approach we need three concepts as follows:

(C1) α is an *individuating concept* (or *individuator*) of x for S at t iff α is a concept that only x satisfies relative to the perspective of S at t (e.g., being to his left).
(C2) Individuator β is *epistemically dependent* on individuator α for S at t iff S at t knows (or believes) that something satisfies β on the basis of knowing (or believing) that something satisfies α and that whatever satisfies α satisfies β, but not vice versa.
(C3) If β is epistemically dependent on α for S at t, and y satisfies β relative to S and t, and x satisfies α relative to S and t, and not-($x = y$), then concerning S's thought [ϕ, β] at t, x is a *deeper* and *more primary* object of that thought by S at t than is y (though y is *also* an object, a relatively superficial, secondary, object of that thought by S at t).

There *is* then a sense in which the subject is "really" reaching for the pepper mill, but a sense acceptable to latitudinarian intuitions. For, relative to that subject and time, and to the situation described by the example, the pepper mill is a "deeper" and "more primary" object of his thought, since the individual concept ⟨the salt-shaker now to my left⟩ is epistemically dependent for *S* at *t* on ⟨the container I saw to my left a while back⟩, or perhaps on ⟨the object with such and such a shape and color that I saw to my left a while back⟩.[5]

C

We have discussed the role of *de re* belief in explanations of intentional action, with examples where such a role seems important and contrary to latitudinarianism. But we have also found simple ways in which latitudinarian intuitions might be protected from such examples. In a new example, the problem is how to explain someone's intentionally taking a basket rather than a dish. That he thinks the basket more valuable and intends to take the more valuable seems insufficient since *under another description* it might be the dish that he then thinks more valuable.[6] So just the fact that under some description he thinks the basket more valuable and wants to take the more valuable is not enough. This has been urged as a problem for latitudinarianism, for it has been thought to prove that *de re* belief about an object requires something *stronger* than just belief about the object under some description (some description that discriminates the object uniquely). Something stronger–yes, but what? For objects in one's environment, it has seemed appropriate to require belief under a mode of perceptual presentation. But this will not do anyhow, since one might believe the dish more valuable under one such mode of presentation while believing the basket more valuable under another.[7] Accordingly, just the fact that under some mode of perceptual presentation the agent thinks the basket more valuable and wants to take the more valuable still falls short.

Such reasoning may lead one to impose a requirement for a satisfactory explanation *E* of *S*'s acting on an object *x*: One might require *E* to identify some *basic* action *B* performed by *S* such that (i) *E* explains why *S* (overridingly) intends to perform *B*, and (ii) *E* also explains how in the circumstances if *S* performs *B* that will bring it about that *x* comes to have property ϕ.

What it is crucial to require concerning *B* specifically is perhaps that it be safe from the threat of "near-contraries," which may be thought of as inconsistent attitudes that someone rational may yet adopt. A case in point is the belief about the dish (under one description or mode of presentation) that it is the more valuable and the belief (under another) that it is not. Analogously one might have the following intentions respecting the dish: (under one description or mode of presentation) to remove it, and (under another) not to remove it.

Thus an explanatory schema for *S*'s removal of a certain object from a place before him might identify a basic action (e.g., reaching-and-grasping) that *S* performs, explain why that agent (overridingly) intends to perform that basic action, and explain also how the circumstances are such that if he does perform it, then as a result the object will be removed by him from the place before him.

What makes such explanation in terms of basic action successful where explanation by *de re* attitudes fails? The key to this may be thought to lie in the fact that basic actions, such as *S*'s reaching-and-grasping, are safe from the threat of near-contrariety. It is presumably for this reason that neither the dish nor the basket can be such that *removing it* is a "basic action." For one can have the intention to-remove-it (under one description or mode of presentation of the object involved) concurrently with the intention *not*-to-remove-it (under another). Since these are near contraries, neither can yield a "basic action," one such that the intention to perform it suffices for its performance.[8]

In view of its importance, this concept of "basic action" deserves a close look. To be safe from the threat of near-contrariety a basic action must admit no incompatible alternative that a fully rational agent could concurrently intend. But there are surely actions not in any reasonable sense "basic" that yet brook no incompatible alternatives and are hence immune to near-contrariety. Consider removing-the-dish. Surely no-one rational could concurrently intend *de dicto* both to-remove-the-dish and *not*-to-remove-the-dish. Indeed, *any* attitude that is like the intention to-remove-the-dish in being (in a sense clear enough here) *de dicto* would repel near-contrariety.

Suppose now an agent in circumstances where if he intends to-remove-the-dish (and is able to do so, etc.) then he does remove-the-dish. Why could we not appeal to this fact along with his actual intention to-remove-the-dish (acquired through his belief that the dish is the more valuable) in explaining this removal of the dish? What sets apart this fact from the fact that if the agent intends to-reach-and-grasp (etc.) then he does reach-and-grasp?

Whatever distinguishes reaching-and-grasping from removing-the-dish so as to make it an appropriate intention–world interface, it seems not to involve resistance to near-contrariety: not anyhow to near-contrariety introduced by indexical or descriptive references other than to oneself as oneself or to the present as the present. We have already noted how removing-the-dish withstands any direct threat of near-contrariety. (No-one fully rational can intend both it and its complement: not-removing-the-dish.) Besides, nothing precludes my intentionally reaching-and-grasping by reaching-and-grasping-with-my-right-hand. Reaching-and-grasping thus seems no more basic than reaching-and-grasping-with-my-right-hand. But *if* the action of removing-the-dish failed to be basic because of its descriptive reference to the dish, which opened it to some indirect threat of near-contrariety, it is plain that reaching-and-grasping-with-my-right-hand could not be secure from such threat and would equally fail to be basic. For the thought content *my right hand* is no less descriptively referential than the thought content *the dish.* So I might intend to reach-with-my-right-hand (under one description of that hand) and not to reach-with-my-right-hand (under another).

Consider then an explanation of one's removal of the dish that gives *removing-the-dish* the status of intention–world interface by simply assuming that one's intention to perform that action will succeed. Is one's simple assumption inevitably doomed by the mere possibility of near-contrariety with respect to such an intention? May it not be true in a certain situation that if the agent had the overriding *de dicto* intention that he remove the dish then he just would remove the dish? As I face a front door it may surely be that if I decided *de dicto* to ring the doorbell in that situation I *would* ring the doorbell (even if my *de re* intention is not there immune to near-contrariety).

What then is wrong with an explanation that sets the intention–world interface at the likes of *removing-the-dish* or *ringing-the-doorbell* by assuming that, in the situation, *de dicto* intention to perform such an action would succeed? So far as I can see, such an explanation makes not an error of commission but at most one of omission. It does not go as deep as it might. Whether or not that is a failing in an explanation, which I doubt, there is anyhow a more complete explanation in the neighborhood. Of course, for *every* explanation, there is always a more complete explanation in the neighborhood. But whether or not the more complete explanation is "better," and the less complete failed, depends on many factors, including the nature of the puzzlement addressed, the time and intelligence available for addressing it, and so forth. Anyhow, is there any *special* (and especially pertinent) sort of incompleteness that pertains to our explanation of intentional action?

Consider again the hierarchy generated by someone *S* who intentionally *A*1's-by-*A*2'ing, intentionally *A*2's-by-*A*3'ing, . . . , and intentionally *An*'s-by-*B*'ing; all while there is no *C* such that he intentionally *B*'s-by-*C*'ing. An explanation for *S*'s *A*1'ing may put the intention–world interface at some *Ai* by simply assuming that in the situation if *S* intends to *Ai* then he does *Ai*. But any such explanation is subject to a kind of incompleteness that pertains especially to explanations of intentional action, and which we may aptly label "intentional incompleteness." For it is an incompleteness owed to the absence from that explanation of attitudes held by the agent which together with the circumstances would explain why he *Ai*'s. Only an explanation that sets the intention–world interface at *B,* the base of our hierarchy, would be intentionally complete. Such an explanation would assume that if the agent intends to perform *B,* then he does perform *B;* and it would explain also how in the circumstances if the agent performs *B* then he thereby performs *A*1.

It is not resistance to near-contrariety that gives a basic action its place as basic, nor does it even seem true that every basic action is in fact safe from near-contrariety in some special way not shared by actions that are not basic. I might intentionally make a signal by raising my right hand in a certain way in the absence of any other action *C* by performing which I intentionally raise my right hand in that way. My raising my right hand in that way would then be basic, or so it does seem. But any *de re* intention on my part regarding my right hand, to raise *it* in a certain way, would be as receptive to near-contrariety as such *de re* intention is ever apt to be. For my right hand might well be such that, under the mode *the hand I feel here,* I intend to raise it, while under the mode *the hand I see there,* I intend to let it lie. (I might through a mirror arrangement confuse my right with my left hand).

The fact that, regarding my right hand, I do under *some* mode of presentation or conception overridingly intend to raise it, while able to do so, does not ensure that I do raise it or even that it does raise. (Recall the mirror case.) Putting aside for now the hard problem of just how to understand ability, on reflection there do seem special modes of presentation or conception under which only is the intention-cum-ability to move a certain limb guaranteed success. (I may be quite able to move my right hand but if I intend to move it only as the hand I see in the mirror, I may succeed in moving nothing or at best my left hand.) It is such *de dicto* intention that bids fair to be truly basic: Only in the carrying out of such *de dicto* intention are we likely to find an action *B* with no deeper *C* such that the agent *B*'s by *C* 'ing.

For Merleau-Ponty we are able "clearly to understand motility as basic intentionality."[9] Here is what he means:

> Movement is not thought about movement, and bodily space is not space thought of or represented. . . . A movement is learned when the body has understood it, that is, when it has incorporated it into its "world," and [we] . . . aim at things through it. (137–9)

> The cultivation of habit is indeed the grasping of a significance, but it is the motor grasping of a motor significance. (143)

> When I put my hand to my knee, I experience at every stage of the movement the fulfillment of an intention which was not directed at my knee as an idea or even as an object, but as a present and real part of my living body. . . . (144–5)

Every such stage of the movement of my hand must then have been intended and thus grasped–at least motor grasped as a motor significance. If it is yet not "thought about movement," not "thought of or represented," this must mean only that I need not have consciously and occurrently intended each such stage separately. (Compare a pianist or typist in furious performance.)

But nor does it seem enough to postulate an intended motion of my hand "as a present and real part of my living body." For a congenital paralytic aided by a contraption may be able to move his hand and intend to move it "as a present and real part of . . . [his] living body." Yet his intention may be quite unlike that of someone ablebodied, though each intends the motion of a hand "as a present and real part" of their own living body. What then is the difference?

For the congenital paralytic his hand is an object that he knows only "from outside." It is a thing at which he can aim through movements of other parts of his body and by means of the contraption. But he may yet aim at it as a certain present and real part of his living body, since it is such a part, and he may surely think of it that way.

Someone ablebodied knows his hand not only from outside but also from within. He may experience every stage of its movement as the fulfillment of an intention. Very well, but just how are such stages represented aforethought?

When I whistle a tune or say a word, my mouth and my tongue in it assume a sequence of positions that I could not spell out in words or picture in a diagram. When typing, fingers strike keys in a sequence one *could* describe beforehand; but doing so word by word would slow typing to a grind.

Yet if someone asks how one whistles the tune or says or types the word, he can easily be offered a demonstration. "Like this," one answers, and gives an instance. There is thus a way of doing what one then does such that (a) one normally does it that way, (b) one knows oneself able to do it that way and able to intend on occasion so to do it, and (c) one's representation of one's *so* doing it is only "from within" (unlike that of the linguist or the music or typing teacher) and expressible–even to oneself–*only* in the actual or mock or imagined doing of it that way.[10]

Having thus approached questions of the phenomenology of the body, of our elusive categories of truly basic action, I regret my lack of space and insight to enter them more deeply at this time. Still, it would be folly to dismiss latitudinarianism just for raising such questions.

D

Let us think of "propositions" as what are expressed by declarative sentences. Latitudinarianism then inclines one to a certain conception of propositions, as follows.

> Suppose several people are disguised blindfolded and are filmed dancing. They later grade their own performances without knowing who is who. When they are identified *S* [the one with the clown costume in the film] complains that the others have given him unjustly low marks, until it is pointed out that he has given *himself* the lowest

marks and in fact has sincerely given himself marks he knew to be lower than those he gave anyone else. . . . If so, then there was someone, himself, that *S* thought was the worst . . . Of course, *S* did not believe of himself as *himself* that he was the worst. That is, he did not believe #I am the worst#.[11]

We thus arrive at the following combination:

(a) $(\exists x)$ [x = *S* & *S* thought *x* was the worst]
(b) Not-(*S* thought #I was the worst#)

Given (a), *S* would seem to have accepted the Russellian, concrete, singular proposition constituted by himself and the property of having been the worst: ⟨*S*, having been the worst⟩. However, given (b), *S* did *not* accept what he could in the context express by the sentence 'I was the worst'. But we are thinking of "proposition" as just "what is expressed by declarative sentences." So there is *some* sort of proposition expressed by S when he affirms "I was the worst," and this expressible that is thus expressed, a "proposition," is one that *S* does *not* accept. Given what we have derived from (a) above, therefore, there is *some* sort of proposition expressed by *S* in saying "I was the worst," some sort of proposition that is *other* than the Russellian, concrete, singular proposition associated with that utterance. What sort of proposition might this be? Let us step back and look at this in perspective.[12]

What is a perspectival proposition? What is a perspective? What is it for a perspectival proposition to be true in a perspective?

The perspectival propositions of interest here are those expressed by first-person, present-tense sentences, with essential indexicals 'I' and 'now' or their equivalents. Such a proposition is not viewed as true or false absolutely, but only as true or false in a context or "perspective" provided by a particular subject and a particular time: Perspectives are such ⟨subject, time⟩ ordered pairs. Such a perspectival proposition expressed by a sentence of the form 'I am now *F*' is true at a perspective iff the subject of the perspective has the property expressed by the predicate '*F*' at the time of the perspective. Note that belief is always someone's belief at a certain time. The perspective of a belief *B* is given by the believer *S* and the time of belief *t*. A belief *B* of a first-person present-tense proposition is true if and only if that proposition *P* is true in the perspective of belief *B*. To believe of oneself as oneself and of the time of belief as the time then present that one is then such-and-such is to believe at that time the first-person, present-tense proposition for the property of being such-and-such.

I may believe that the man I see (in a mirror, myself as it happens) has his pants on fire, and so believe about someone who is none other than myself that his pants are on fire; all without yet believing that *I* have my pants on fire. And only when I believe the latter do I roll on the ground instead of just shouting at the man I see. Consider then the two beliefs:

(B1) that *the man I see* is on fire;
(B2) That *I* am on fire.

Even though each of these attributes the same property to the same man, only the second plays a certain crucial role in explanation of that man's action. Thus the indexical '*I*' in the expression of B2 seems "essential": Replacing it with a co-designating singular term, 'the man I see', entails a crucial loss of explanatory power. This poses the "problem of the essential indexical."

Here is John Perry on that problem:

I argue that the essential indexical poses a problem for various otherwise plausible accounts of belief. I first argue that it is a problem for the view that belief is a relation between subjects and propositions conceived as bearers of truth and falsity. The problem is not solved merely by replacing and supplementing this with a notion of *de re* belief. *Nor is it solved by moving to a notion of proposition which, rather than true or false absolutely, is only true or false at an index or in a context C (at a time, for a speaker, say).*[13]

Thus Perry argues against the very conception of proposition that we have been led to by latitudinarian intuitions. What is his argument? Here is the key passage:

I say "I believe that I am making a mess." On our amended doctrine of propositions, this ascribes a relation between me and *that I am making a mess*, which is a function from indices [or contexts, or perspectives] to truth values. The belief report seems to completely specify the relativized proposition involved; there is no missing conceptual ingredient. So the problem must be solved.

But it is not. I believed that a certain proposition, *that I am making a mess,* was true–true for me. So belief that this proposition was true for me then does not differentiate me from the other shopper [who also believes that that proposition is true for me], and can not be what explains my stopping and searching my trolley for the torn bag. . . .

The fact that I believed this proposition true for Perry at the time he was in the supermarket does not explain my behavior either. For so did the other shopper.[14]

This line of reasoning at bottom assumes that believing a perspectival proposition P in a perspective $\langle S, t\rangle$ is equivalent to believing that P is true in $\langle S, t\rangle$. But this assumption is false, for several reasons. For example, suppose I believe that the first-person proposition *that I am on fire* is true in the perspective constituted by the man I see and the present time. So about that perspective ⟨the man I see, the present time⟩ I believe that the first-person proposition *that I am on fire* is true in that perspective. But, even though, unbeknownst to me, I am the man I see, I still do *not* believe at that time the first-person proposition itself, *that I am on fire*. On certain easily granted, harmless assumptions (e.g., that I speak only English), only when I am prepared to say "*I* am on fire" do I believe that first-person proposition itself, *that I am on fire.*

Frege, too, like Perry, rejects perspectivism unequivocally.[15] But to my knowledge there is in his writings no explicit argument for the assumption that truth must be absolute, unless we count the following as an argument:

> What if beings were even found whose laws of thought flatly contradicted ours and therefore frequently led to contrary results even in practice? The psychological logician could only acknowledge the fact and say simply: Those laws hold for them, these laws hold for us. I should say: We have a hitherto unknown type of madness. Anyone who understands laws of logic to be laws that prescribe the way in which one ought to think–to be laws of truth, and not natural laws of human beings' taking a thing to be true–will ask, who is right? Whose laws of taking-to-be-true are in accord with the laws of truth? The psychological logician cannot ask this question; if he did he would be recognizing laws of truth that were not laws of psychology. One could scarcely falsify the sense of the word "true" more mischievously than by including in it a reference to the subjects who judge. Someone will now no doubt object that the sentence "I am hungry" can be true for one person and false for another. The sentence, certainly–but not the thought; for the word "I" in the mouth of the other person denotes a different man, and hence the sentence uttered by the other person expresses a different thought. All determinations of the place, the time, and the like, belong to the thought whose truth is in point; its truth itself is independent of place or time. How, then, is the Principle of Identity really to be read? Like this, for instance: "It is impossible for people in the year 1893 to acknowledge an object as being different from itself"? Or like this: "Every object is identical with itself"? The former law concerns human beings and contains a temporal reference; in the latter there is no talk either of human beings or of time. The latter is a law of truth, the former a law of people's taking-to-be-true. The content of the two is wholly different and they are independent of one another; neither can be inferred from the other. Hence it is extremely confusing to designate both by the same name, "Principle of Identity." These mixings-together of wholly different things are to blame for the frightful unclarity that we encounter among the psychological logicians.[16]

One way of reading this passage has Frege arguing in effect that perspectivism involves psychologism and must be rejected for that reason. Although this reading may seem strained, the passage does suggest that perhaps Frege convicted perspectivism because he saw it as too closely associated with abominable psychologism. But it is far from clear that the two *are* too closely associated. In fact, the thought that they are may just conceivably derive from a confusing ambiguity in

(i) X is true for S at t

between

(ii) At t, S takes X to be true

and

(iii) X is true in the perspective $\langle S, t, \rangle$.

Only (ii) is directly relevant to Frege's hated psychologism. But since (ii) and (iii) are logically quite independent–neither entails the other–it is hard to see what significant connection can be found here between psychologism and perspectivism.[17]

Frege's objectivism is not an unargued dogma. His detailed critique of subjectivist psychologism is legendary for its subtlety and power. But Frege's objectivism rejects not only psychologism but also perspectivism. And the guilt he pins on the latter is at most its association in his mind with the former.

Frege is not alone, however, in assuming without argument the absolutism of truth. Nearly all who postulate propositions as objects of attitudes share such an assumption. Not that it is inevitably a shame to make an unargued assumption, so long as there is no specific reason to suspect it, and so long as it coheres well and is simple by comparison with equally coherent possible alternatives. But unlike Frege we have now seen how our consciousness of the self and the present gives reason to question absolutism. And I know of no countervailing and compelling argument to prefer absolutism to perspectivism.

Latitudinarianism, explained at the outset by LB and LI, may also be described more briefly and sketchily as the doctrine that reference is always through an individuating concept, that thought about an entity is always "under a description" or anyhow "under an individuating concept" which uniquely specifies that entity. In this paper I have given latitudinarianism a less linguistic form than in the past, and I have tried to defend it from several objections. This defense has revealed the natural alliance between such latitudinarianism and a view of concepts and propositions as (at least sometimes) perspectival rather than (always) absolute: that is, as satisfied (concepts) or true (propositions) not absolutely but in a perspective. I have tried to defend this alliance from several objections, and also to highlight some of its advantages.[18]

NOTES

1. A more linguistic version of latitudinarianism is defended in my "Propositional Attitudes *De Dicto* and *De Re*," *Journal of Philosophy* 67 (1970): 883–96," and my "Rejoinder to Hintikka," *Journal of Philosophy* 68 (1971): 498–501.
2. This problem is drawn from Richard Feldman's "Actions and *De Re* Beliefs," *Canadian Journal of Philosophy* 8 (1978): 577–82. Cf. E. Sosa and M. Pastin, "A Rejoinder on Actions and *De Re* Belief," *Canadian Journal of Philosophy* 11 (1981): 734–9.
3. This is to suggest only that despite x and y being *indistinguishable* from the perspective of S's *de re* beliefs (every *de re* belief by S about x having a correlate

de re belief by *S* about *y*), there may yet be important differences between *x* and *y* via-à-vis *S*, since for one thing some important *de re* belief of *S*'s about *x* may be *true* while the corresponding belief by *S* about *y* is *false*; and some such difference may still help explain why *S* acts on *x* and not on *y*. But this is *not* to suggest that *only* one's true *de re* beliefs about an object can help explain one's actions on that object. Indeed we shall encounter a counterexample below: the example of the dish and the basket, where the agent selects the basket through believing it (*falsely*) to be more valuable.

4. This problem is drawn from Thomas McKay, "Actions and *De Re* Beliefs," *Canadian Journal of Philosphy* 14 (1984): 631–5.
5. The proposed solution is due materially to David Sosa. Other puzzles seem also amenable to this approach: for example, the martini-drinker case and other cases in Keith Donnellan's "Reference and Definite Descriptions," *Philosophical Review* 75 (1966): 281–404. It is a solution "in a latitudinarian spirit" since it does not explicate reference of any sort through any brute causal relation, but insists that reference, even deep and primary reference, is always "under a description" or "through an individuating concept or sense." For this "broad-gauge latitudinarianism," then, belief under a description or through an individuating concept (*any* such description or concept) is *both* necessary *and* sufficient for "superficial" or "secondary" reference, *and* belief under a description or through an individuating concept is *necessary* even for "deep" or "primary" reference.
6. The subject might believe, first, *that the basket is the more valuable* and at the same time believe, second, *that the more valuable is the more valuable*; and it might just be that, despite his first belief, it is really the dish that is the more valuable. This example is due to Herbert Heidelberger, "The Self-Presenting," *Grazer Philosophische Studien* 7 (1979): 59–76: 76.
7. This may be seen from the argument against narrower ("classical") conceptions of *de re* belief in Lynne Rudder Baker's "*De Re* Belief in Action," *Philosophical Review* 91 (1982): 363–7.
8. This is the approach defended by Baker, "*De Re* Belief in Action."
9. Maurice Merleau-Ponty, *The Phenomenology of Perception*, trans. C. Smith (New York: Humanities Press, 1962), p. 137. (Page references in parentheses in the text will be to this work.)
10. I am told that psychologists call such knowledge "procedural."
11. "Propositional Attitudes *De Dicto* and *De Re*," p. 853; letting the #'s take the place of the corner quotes used in the article, '*S* believes #α is *F*#' is explained on p. 885 as "*S* has a belief (in a proposition) that, given normal circumstances, he could correctly express in our language by asserting the sentence composed of α followed by 'is' followed by '*F*' (where '*F*' is taken not as a metalinguistic variable but as a predicate of the language).
12. The conception of propositions thus adumbrated in the passage just quoted is developed further in my "Propositions and Indexical Attitudes," in Herman Parret (ed.) *Believing: Epistemological and Semiotic Approaches* (Berlin and New York: Walter de Gruyter, 1983), pp. 316–32. It is unusual, but others have also adopted such a conception of "propositions": e.g., Bernard Williams, in *Descartes: The Project of Pure Enquiry* (Penguin, 1978); see esp. Appendix 1. And others have also presented it, if only to argue against it: e.g., John Perry, in "The Problem of the Essential Indexical," *Noûs* 13 (1979): 3–21; reprinted in Nathan Salmon and Scott Soames, *Propositions and Attitudes* (Oxford: Oxford University Press, 1988); pp. 83–102 (see esp. pp. 92–5); references here will be to this reprinting.

13. Perry, "The Problem of the Essential Indexical," p. 84. Emphasis added. The problem was brought to prominence by Perry and, earlier, as Perry reports, by Hector Castañeda. See Castañeda's "'He': A Study in the Logic of Self-Consciousness," *Ratio* 8 (1966): 130–57. See also several of the essays and replies in *Agent, Language, and the Structure of the World: Essays Presented to Hector-Neri Castañeda, with His Replies*, ed. James E. Tomberlin (Indianapolis: Hackett, 1983).
14. Ibid., pp. 93–4.
15. Gottlob Frege, "The Thought: A Logical Inquiry," trans. A. M. and Marcelle Quinton, In *Philosophical Logic*, ed., P. F. Strawson (Oxford: Oxford University Press, 1967).
16. Gottlob Frege, *The Basic Laws of Arithmetic*, trans. and ed., with an introduction, by Montgomery Fürth (Berkeley and Los Angeles: University of California Press, 1964), pp. 14–15.
17. Interestingly, one might find the very same confusion at the core of Perry's argument against perspectivism discussed above.
18. I have been helped by the works cited above, and by discussion with David Martens and David Sosa. My thanks also to Diana Raffman and Walter Sinnott-Armstrong.

15

T-Sentences

SCOTT SOAMES

What is it for a definition of truth for a language to be materially adequate? The easy answer would seem to be that a definition of truth is materially adequate if the *definiens* applies to all and only the true sentences of the language. But as a criterion of adequacy this may not seem helpful, for one may wonder how we are to determine in a given case if the definiens really does apply to precisely the truths. We can thank Alfred Tarski for suggesting an apparently sure-fire way of putting this worry to rest. The definiens of a proposed definition will apply to precisely the truths if for each sentence of the object language, an instance of Schema T,

Schema T: s is true in L iff p

follows logically from the definition – where an instance is obtained by replacing 's' with a name (e.g., a quote-name) of a sentence of the object language L and replacing 'p' with a metalanguage sentence that means the same as that object-language sentence. If this criterion is met, then for any object-language sentence s, the conditions for establishing that the definiens applies to it will be the same as the conditions for establishing some sentence that means the same as s, and hence for establishing s itself.

Although this paper does not comment on the work of Ruth Marcus, its roots, along with that of all my work on truth, are directly traceable to her. In the late 1970s, when I was a beginning assistant professor at Yale, I realized that I needed to learn something about the notion of truth, and so decided to teach a graduate seminar on the topic. I consulted with Ruth about the readings for the course, and she suggested some material that I was not familiar with at the time. She also attended a number of the seminars, along with another of our colleagues, Robert Fogelin. I benefited greatly from their support and from their comments. Ruth, in particular, urged me to publish the seminar presentations as a book, but I resisted, being all too aware of their shortcomings, and of my continuing ignorance about important parts of the subject. She did, however, interest a publisher, and got me to commit myself to a long term project of doing a book on truth. None of us realized at the time how long term the project would be, or that I would still be working on it, along with too many other things, today. I can only hope that when the project is finally finished, it will vindicate her trust. Over the years Ruth has been a dear friend, a model, and an inspiration to me; no one has been more important in my professional life.

If we can show that a truth definition meets this criterion, not only will our definiens be a truth predicate, it will provably be a truth predicate. Why are we so confident of this? Well, it might be answered, we are confident because it is part of our very concept of truth, of what we mean by the predicate *true*, that the claim that a sentence is true is materially equivalent to the sentence itself (and hence to any sentence with the same meaning). For surely, if a sentence is true, then the claim that it is true must also be true, and vice versa. In light of this, it may seem that the inference from a sentence to the claim that it is true (and vice versa) is bound to be acceptable. For example, someone who accepted the English sentence *Snow is white* while denying that it is true, or who accepted the claim that the sentence is true while failing to accept the sentence itself, might well be taken to have shown that either he didn't understand the sentence or he didn't understand the truth predicate. A corresponding point might be made about homophonic instances of Schema T–that is, instances of Schema T in which the sentence replacing 'p' is the very sentence whose quote name replaces 's' in the schema. Putting aside problematic cases such as those involving context sensitivity, indeterminacy, and sentences containing semantic terms, it might plausibly be maintained that it is part of our very concept of truth, and of what we mean by the predicate *true*, that homophonic instances of Schema T are true, and assertable, and that someone who denied such an instance could be presumed not to understand it. Moreover, if homophonic instances are true and assertable, then translations of them should also be true and assertable. In particular, instances of Schema T in which 'p' is replaced with a sentence that is a translation of the sentence whose quote-name replaces 's' should be true, and assertable.

Suppose, now, that we think of a definition of truth for a particular language in the way I suggest Tarski did–namely as a stipulative definition that introduces a new, restricted truth predicate designed to do the theoretical work, over a limited domain, of our unrestricted, pretheoretic notion of truth. Suppose further that our definition is materially adequate in Tarski's sense. What could we conclude from this about the relationship between the new predicate introduced by the definition and the one it is supposed to replace? As we have seen, there is no problem in concluding that the two are coextensive over the object language. Still, it is far from clear that mere coextensiveness is enough to guarantee that the new truth predicate will be sufficient for doing the theoretical work of our ordinary notion.

We would have such a guarantee if we could be assured that our definition had captured the very meaning of our ordinary truth predicate, or at least had captured the central core of its meaning when the predicate is understood as restricted to a well-behaved object language. But now, one might wonder, doesn't material adequacy in Tarski's sense provide just such a guarantee? For if instances of Schema T really do follow logically from the stipulative definition, then these instances will be analytic, if anything is. Certainly a stipulative definition is analytic, if anything is; and anything that follows logically from something analytic is itself analytic. Moreover, if it really is part of what we mean pretheoretically by the word *true* that all homophonic instances of Schema T are both true and assertable, then it would seem that homophonic instances of Schema T are also analytic. Indeed, we might well speak of such T-sentences as "partial definitions" of what we mean by the ordinary predicate *true*. Finally, if coming to understand the ordinary predicate amounts essentially to learning this, that is, to learning all these "partial definitions" or something that entails them, then it might well seem that the new predicate introduced by the definition captures not just the extension, but also the intension, or meaning, of the ordinary one.

This, it seems to me, is a rather seductive line of reasoning, one that, I suspect, has influenced a number of philosophers, including Carnap and, in certain moments, even Tarski. Unless I miss my guess something very much like it stands behind Tarski's references to homophonic instances of Schema T as "partial definitions" of truth,[1] as well as his remark in "The Semantic Conception of Truth and the Foundations of Semantics," that his definition "does not aim to specify the meaning of a familiar word used to denote a novel notion; on the contrary, it aims to catch hold of the actual meaning of an old notion."[2] I suspect further that this line of reasoning stands behind his confidence, expressed later in the same article, that his formally defined semantic notions are capable of playing a central role in defining *meaning* and *synonymy*, and in developing theories of meaning for natural languages.[3] This confidence is understandable if one recognizes the obvious intensional connection between the ordinary notions of truth and meaning, while at the same time taking Tarski's materially adequate definition of truth in L to have captured the meaning of the ordinary truth predicate (restricted to L).

What I mean by the intensional connection between the ordinary notions of truth and meaning is that the two concepts are related so as to make instances of (1)

1. If s means in L that p, then s is true in L iff p

trivial a priori truths, which can be established by simple reflection, without empirical investigation.[4] Given this minimal intensional connection between the two notions, one can use statements of truth conditions to draw certain minimal conclusions about meaning. For example, suppose I establish, either empirically or by being told, something of the form (2).

2. s is true in L iff p.

Since instances of (1) are obvious a priori truths, I can use them, without further investigation, together with my claim of the form (2), to conclude, for example, something of the form *s does not mean in L that ~ p*; for if s did mean this, then by (1) we would have the relevant instance of *s is true in L iff ~ p,* which contradicts our initial instance of (2). The intensional connection between truth and meaning may be stronger than this, but it must be at least this strong if statements of truth conditions are going to tell us anything about meaning. Given the belief that statements of truth conditions do tell us something about meaning, and also the conviction that a materially adequate definition captures the meaning, or intension, of the ordinary truth predicate, and hence the intensional connection between truth and meaning, one can understand the confidence some philosophers have had that Tarksi's formally defined truth predicate could be used to play a central role in the study of meaning.

As I have stressed, the line of reasoning leading to the conclusion that a materially adequate truth definition captures the meaning of the ordinary truth predicate, when it is understood as restricted to a well-behaved object language, is highly plausible. However, plausibility is one thing, truth another. For it is quite clear that the meaning, or intension, of the definiens provided by Tarski is very different from that of our ordinary truth predicate. This difference can be seen by considering a very simple artificial language L containing one predicate, *W,* one name, *s,* one atomic sentence, *Ws,* one variable, *x,* plus negation, conjunction and quantification. A Tarskian definition of truth for this language is given by (3).

3a. Truth relative to an object (i.e., a one-member sequence)[5]
For all formulas F of L and objects o, F is true relative to o iff there is a set T_s of which ⟨F, o⟩ is a member and for all formulas G of L and objects o′, ⟨G, o′⟩ is a member of T_s iff (a)G = *Ws* and *W* (which applies to an object iff it is white) applies to the denotation of *s* (which denotes snow) (i.e., G = *Ws* and snow is white), or G = *Wx* and *W* applies to the denotation of *x* relative to o′ (i.e., o′ is white); or (b) G = ***(~A),*** for some formula A of L, and ⟨A, o′⟩ is not a member of T_s; or (c) G = ***(A & B),*** for some formulas A and B of L, and ⟨A, o′⟩ and ⟨B, o′⟩ are both members of T_s; or (d) G = ***(∃ x A(x)),*** for some formula A(x) of L, and there is an object o* such that ⟨A(x), o*⟩ is a member of T_s.[6]

b. Truth
For all sentences S of L, S is true iff for all objects o, S is true relative to o.

Tarski's truth predicate, the definiens of his definition, is given in (4).[7]

4. For all objects o, ⟨S, o⟩ is a member of a set T_s such that for all formulas G of L and objects o′, ⟨G, o′⟩ is a member of T_s iff (a) G = *Ws* and snow is white or G = *Wx* and o′ is white; or (b) G = ***(~A),*** for some formula

A of L, and ⟨A, o′⟩ is not a member of T_s; or (c) G = ***(A & B)***, for some formulas A and B of L, and ⟨A, o′⟩ and ⟨B, o′⟩ are both members of T_s; or (d) G = ***(∃ x A(x))***, for some formula A(x) of L, and there is an object o* such that ⟨A(x), o*⟩ is a member of T_s

The striking difference in meaning between Tarski's truth predicate, (4), and our ordinary truth predicate can be seen by comparing the T-sentence (5), which contains our ordinary truth predicate, with the T-sentence (6), which contains Tarski's truth predicate.

5. *Ws* is true in L iff snow is white.
6. [For all objects o, ⟨*Ws*, o⟩ is a member of a set T_s such that for all formulas G of L and objects o′, ⟨G, o′⟩ is a member of T_s iff (a) G = *Ws* and snow is white or G = *Wx* and o′ is white; or (b) G = ***(~A)***, for some formula A of L, and ⟨A, o′⟩ is not a member of T_s; or (c) G = ***(A & B)***, for some formulas A and B of L, and ⟨A, o′⟩ and ⟨B, o′⟩ are both members of T_s; or (d) G = ***(∃x A(x))***, for some formula A(x) of L, and there is an object o* such that ⟨A(x), o*⟩ is a member of T_s] iff snow is white.

We can reduce the clutter in (6) by noting that in this case only clause (a) in the characterization of the set T_s is relevant to the object-language sentence *Ws*. Thus, the information provided by (6) about this sentence is essentially the same as that provided by (7).

7. Snow is white iff [for any o, ⟨*Ws*, o⟩ is a member of a certain set T_s, which contains it iff snow is white].

The point to notice is that (6) and (7) tell us nothing about the meaning of the object-language sentence *Ws*. If one doesn't know the meaning of this sentence, and is then given the information in (6) and (7), one can draw no conclusion whatsoever about what *Ws* does or does not mean. For example, one cannot conclude that *Ws* does not mean that snow is black, or even that snow is not white. By contrast, the T-sentence (5), containing the ordinary truth predicate, does tell us something about the meaning of *Ws*. In particular, if one doesn't know the meaning of this sentence, and then is given the information in (5), one can draw the conclusion that *Ws* does not mean that snow is black, or that snow is not white. Because certain conclusions about meaning can be drawn from the information provided by (5) but cannot be drawn from the information provided by (6) and (7), the information provided by (5) is not the same as the information provided by (6) and (7). In short, the ordinary truth predicate differs in meaning from Tarski's truth predicate, and because of this, T-sentences containing the ordinary truth predicate differ in meaning from T-sentences containing the predicate provided by Tarski's definition.

What is it about the meaning of the ordinary truth predicate that allows us to derive certain weak conclusions about meaning from T-sentences containing it which is lacking in the meaning of Tarski's formal substitute? It is just the connection between truth and meaning expressed by (1) mentioned earlier. Because of what we mean by ***true*** and ***means,*** instances of (1) are trivial a priori background assumptions that are always available to be combined with T-sentences containing the ordinary truth predicate to draw conclusions about the meaning of object language sentences. The same is not true of instances of (8), which results from (1) by substituting Tarski's truth predicate for our ordinary one.

8. If s means in L that p, then ([for all objects o, ⟨s, o⟩ is a member of a set T_s such that for all formulas G of L and objects o′, ⟨G, o′⟩ is a member of T_s iff (a) G = *Ws* and snow is white or G = *Wx* and o′ is white; or (b) G = ***(~A),*** for some formula A of L, and ⟨A, o′⟩ is not a member of T_s; or (c) G = ***(A & B),*** for some formulas A and B of L, and ⟨A, o′⟩ and ⟨B, o′⟩ are both members of T_s; or (d) G = ***(∃x A(x)),*** for some formula A(x) of L, and there is an object o* such that ⟨A(x), o*⟩ is a member of T_s] iff p)

Unlike instances of (1), it is not the case that arbitrary instances of (8) are trivial and a priori; hence they are not automatically accessible for the drawing of conclusions about meaning from T-sentences like (6) or (7).[8]

In light of this there should be no question that Tarski's definition does not capture the meaning of the ordinary notion of truth, even when that notion is restricted to a particular language. Thus, Tarski's occasional remarks to the contrary are mistaken. How important this is to his conception of his project is a more complicated matter. In my view, Tarski's primary philosophical purpose was to provide formally defined truth predicates capable of fulfilling all the theoretical functions for which we need a notion of truth, while avoiding the difficulties and paradoxes that plague our ordinary notion. Although Tarski's formally defined substitutes are theoretically fruitful in important ways, they clearly cannot be used to give interpretations to formal languages, or to construct theories of meaning and understanding for natural languages. Thus, if we need a notion of truth for these purposes, then we need a notion of truth closer in meaning to the ordinary one than Tarski was able to provide.[9]

That, however, is not the point I wish to pursue here. Rather, I want to look again at the line of reasoning that led to the mistaken conclusion that Tarski had captured the meaning of our ordinary notion. That argument was roughly this:

(i) (Putting aside problematic cases such as those involving context sensitivity, indeterminacy, and sentences containing semantic terms) it is part of what we mean by the predicate *is true* that instances of Schema T – *s is true in L iff p* – are true and acceptable (where an instance is the result of replacing 'p' with a sentence that means the same as the sentence of L whose name replaces 's'). Anyone who knows what *is true* means knows that this is so. Further, it is part of what we mean by *is true* that homophonic instances of Schema T – *'S' is true in English iff S* – are true and acceptable. Such homophonic instances are analytic, as is shown by the fact that anyone who sincerely doubts them must fail to understand them.

(ii) T-sentences that follow from a stipulative definition of a truth predicate that is materially adequate in Tarski's sense are analytic (since they follow from a stipulative definition that is itself analytic).

(iii) Any definition that captures the analyticity of T-sentences has captured the central core of the meaning of the ordinary truth predicate.

We have already seen that these premises lead to a false conclusion. Our next task is to determine where the argument goes wrong.

One point to be noted right away concerns premise (ii). Strictly speaking, T-sentences like (6) are not logical consequences of Tarski-style truth definitions. Rather, they are logical consequences of such definitions plus some very elementary set theory and syntax. Thus, we cannot show that these T-sentences are analytic by noting that stipulative truth definitions are analytic, and that anything that logically follows from something analytic is also analytic. However this is only a quibble. If premises (ii) and (iii) were modified to represent a liberalized conception of analyticity, according to which something is analytic if it follows from a stipulative definition plus elementary set theory and syntax, we would still reach an obviously false conclusion from initially plausible premises. So this can't be the fundamental place where the argument goes wrong.

More interesting issues are raised by premise (i). It has, I think, sometimes been regarded as a vulgar mistake to think that homophonic instances of Schema T are analytic; and, on some conceptions of analyticity, that may be so. However, there are other interesting and natural conceptions of analyticity in which the issue is more delicate. Two such conceptions are A_1 and A_2.

A_1 s is analytic in L iff being a competent speaker of L who understands s (with the meaning it has in L) is sufficient for accepting it and concluding that it is true (in one's language).

A_2 s is analytic in L iff understanding s (with the meaning it has in L) is sufficient for believing the proposition it expresses in L.

English sentences that have been held to be analytic in these senses include (9–11).

9. Bachelors are unmarried.
10. It is not the case that both snow is white and snow is not white. ~(A & ~A)
11. If snow is green, then either snow is green or grass is white. A ⊃ (A v B)

The last two examples were chosen with W. V. Quine in mind. In *Word and Object,* he contends that objective constraints on the translation of truth-functional connectives preclude the possibility that any speakers could correctly be characterized as accepting sentences translatable in the form *A and not A.*[10] Similarly, in *The Roots of Reference,*[11] he considers the doctrine that the truths of logic are true in virtue of the meanings of the logical words, and therefore are analytic. He says:

> But now in terms of the learning process can we perhaps find some sense for the doctrine? We learn the truth functions, I just now suggested, by finding connections of dispositions, e.g. that people are disposed to assent to an alternation when disposed to assent to a component. The law that an alternation is implied by its components is thus learned, we may say, with the word 'or' itself, and similarly for the other laws.[12]

And again:

> Language is social, and analyticity, being truth that is grounded in language, should be social as well. Here then we may at last have a line on a concept of analyticity: a sentence is analytic if everybody learns that it is true by learning its words.[13]

I mention these remarks of Quine not to endorse his view of which sentences are analytic in this sense, or his explanation of how we should construe the meanings of any particular words or sentences, but simply to indicate that we have here a conception of analyticity that may be favorably received even in the most skeptical quarters.

One word of qualification about this conception should be noted at the outset. As they stand, A_1 and A_2 are highly restricted. For example, although they arguably apply to certain very simple logical truths, such as (10) and (11), they do not apply to more complex examples recognition of whose status as logical truths requires not just understanding but also sustained reasoning. One way of remedying this would be to classify as analytic all sentences that are logical consequences of sentences satisfying A_1, or A_2. However, this raises issues about whether we want all of logic to count as being true in virtue of meaning, or whether some disputed principles, such as the law of the excluded middle, should be given a different status, allowing for genuine disagreement (for example between classical logicians and intuitionists) without divergence in meaning. Fortunately for my purposes, it is not necessary to resolve this issue. The sentences containing the ordinary truth predicate that I wish to examine are simple enough so that the question of whether they are analytic can be discussed in terms of A_1 and A_2.

Examples of sentences containing the truth predicate that seem clearly to be analytic are those in (12).

12a. It is true that the earth is round iff the earth is round.
   b. The proposition that the earth is round is true iff the earth is round.

To show that these are analytic we observe that someone who knows what the word *true* means will know that instances of

13a. (It is true that s) iff s
   b. The proposition that s is true iff s

are acceptable, and true. In the case of any particular instance, such as (12a) or (12b), a speaker who understands the sentence that replaces 's' will accept the resulting instance and regard it to be true. Hence, the sentences in (12) are analytic in the sense of A_1. To show that they are analytic in the sense of A_2, we appeal to the weak disquotation principle D.

D. If a sincere, reflective, and competent speaker *x* of English understands and accepts an indexical free sentence s of English, and believes (knows) it to be true (in his language), then x satisfies ***y believes (knows) that s***.

Given that the speaker understands the sentences in (12), while accepting them, and believing them to be true, we can use D to conclude that the speaker believes the propositions expressed by these sentences.

What about homophonic instances of Schema T, like those in (14)?

14a. 'The earth is round' is true in English iff the earth is round.
   b. 'Snow is white' is true in English iff snow is white.

Are they analytic?

If one compares these sentences to those in (12), it might seem as if there is an obvious difference between the two. Whereas the sentences in (12) are clearly necessary, those in (14) are commonly regarded as expressing contingent facts about English. Suppose, for the sake of argument, that we grant this. On some conceptions of analyticity, the contingency of the sentences in (14) would be enough to show that they are not analytic. However, it is not enough to show that they are not analytic in the sense of A_1 or A_2; for these characterizations do not mention necessity, and it is not obvious that only necessary truths can satisfy them.[14]

Indeed, it would seem that the same sort of argument that led us to say that the sentences in (12) are analytic will also lead us to say that the sentences in (14) are. First, paradoxical, context-sensitive, and other problematic sentences aside, a speaker of English who knows what the construction ' ' *is true in English iff* means will know that all instances of Schema T_E,

Schema T_E: s is true in English iff p

are true and acceptable, where an instance is the result of replacing 's' with a quote name of a declarative sentence of English and replacing 'p' either with that very sentence, or with a sentence that means the same as it does. Second, in the case of a particular homophonic instance, such as those in (14), an English speaker who understands the sentence replacing 'p' as well as the quote name of that sentence, will understand the instance, accept it, and believe it to be true. Weak disquotation will then establish that the speaker believes the proposition it expresses. Hence, it seems that unproblematic, homophonic instances of Schema T_E, like those in (14), are analytic in the sense of A_1 and A_2.

In fact, it would appear that we can go further. If examples like those in (14) are analytic, then shouldn't knowledge of that which they express be a priori, in the traditional sense expressed by AP?

AP A proposition p is knowable a priori iff it is possible to come to know *p* simply by entertaining and reflecting upon it, without relying on empirical evidence for justification or verification.

In applying this conception of *aprioricity* to the propositions expressed by T-sentences, we need to keep an important distinction to keep in mind. According to AP a proposition p is a priori iff there is a way of entertaining it such that, by entertaining it in that way, and reasoning about it, one can come to know it. It is not required that every way of entertaining p be such that by entertaining p in that way, and reasoning about it, one can come to know it. This distinction will be significant if there are different ways of entertaining the same proposition, such that entertaining it one way leads one to accept it without appeal to empirical evidence, whereas entertaining it another way does not.

For example, suppose that translation of a sentence preserves the proposition it expresses, and also that one can entertain the proposition expressed by a sentence either by understanding the sentence itself, or by understanding one of its translations. Where p is the proposition expressed by an instance of Schema T, such as (14b), there are clearly ways of entertaining p such that one cannot come to know p simply by entertaining it that way and reasoning about it. For example, if one is a monolingual Spanish speaker who is able to entertain p only by virtue of understanding some translation (14c),

14c. 'Snow is white' es verdadero en ingles si y solo si la nieve es blanca

of (14b), then reasoning alone will provide no basis for taking p to be true. Suppose, on the other hand, that one is an English speaker who is able to entertain p simply by virtue of understanding (14b). If, as we have argued, this sentence is analytic in the sense of A_1 and A_2, then it would seem that entertaining p in this way allows one to conclude that it is true, and hence to come to know it, without appeal to empirical evidence for justification. If so, then instances of Schema T are a priori.

One response to this surprising and somewhat disturbing conclusion might be to maintain that we have the wrong characterization of aprioricity. What we need, it may be suggested, is a conception that will count a proposition as being a priori only if every way of entertaining it is such that, by entertaining it in that way, and reasoning about it, one may come to know it. The idea behind this suggestion would be to grant that there are different ways of entertaining the same proposition, including understanding intertranslatable sentences, but to reserve the label 'a priori' for propositions knowable independent of experience, no matter how they are grasped.

The problem with this proposal, in my view, is that it would exclude a great many propositions that traditionally, and properly, have been regarded as a priori. For example, imagine a slight extension of Saul Kripke's case of puzzling Pierre.[15] Pierre, you will recall, is a bilingual speaker of French and English who does not realize that *London* is the English translation of the French word *Londres*, or even that the two are coreferential. Because of this, he assents to *Londres est jolie*, while dissenting from *London is pretty*. Since these sentences are translations of one another, they express the same proposition. Hence, he accepts a proposition when it is presented via one sentence, while rejecting it when it is presented via another sentence.

Let us now imagine that there are a number of Frenchmen in London in the same linguistic and epistemic situation as Pierre. When they are together, they converse in French–standard French plus one addition. Since they are unaware that *Londres* names the city they live in, they use the name *London* for that purpose. Thus, in their dialect of French, the sentences

15a. Londres est jolie

and

15b. London est jolie

express the same proposition (even though Pierre and his friends are not aware that they do). Now consider a material conditional

16. Si Londres est jolie, London est jolie

with (15a) as antecedent and (15b) as consequent. Although (16) expresses the same proposition as its English translation,

17. If London is pretty, London is pretty

Pierre can come to know this proposition, without appealing to empirical evidence, when it is presented via (17), but not when it is presented via (16). Surely, however, we don't want to deny, on this account, that one can know a priori that if London is pretty, then London is pretty. Thus, we don't want a conception of *aprioricity* that counts a proposition as a priori only if every way of entertaining it is such that, by entertaining it in that way, and reflecting about it, one can come to know it without appeal to empirical evidence.

This argument can be generalized in a variety of ways, provided that we make three assumptions: First, attitude ascriptions report relations to propositions; second, substitution of expressions with the same meaning, or content, in a sentence preserves the proposition expressed by the sentence; and third, it is possible to understand each of a pair of expressions which have the same meaning, or content, without realizing that they do. Persuasive examples of the latter have been given by Saul Kripke *furze/gorse*,[16] Nathan Salmon *catsup/ketchup*,[17] and Stephen Reiber *wager/bet*.[18] Although it is plausible to suppose that these are examples of genuine synonyms, in each case it is possible for a competent speaker to learn them separately, understand them correctly, and yet not be sure that they apply to exactly the same things. In such cases, there will be pairs of sentences

18a. $\forall x\ (Fx \equiv Fx)$
b. $\forall x\ (Fx \equiv Gx)$

which express the same proposition, even though one of these sentences is such that understanding it provides one with a sufficient basis for accepting it, and the proposition it expresses, without appeal to empirical evidence; whereas understanding the other sentence does not. Since we don't want to deny, on this account, that the proposition expressed by (18a) is a priori, we don't want a conception of aprioricity that counts a proposition as a priori only if every way of entertaining it is such that, by entertaining it in that way, and reflecting about it, one can come to know it, without appeal to empirical evidence.

This means that we still have no answer to the surprising and disturbing conclusions that homophonic instances of Schema T are analytic, and that the propositions they express are a priori. However, it does not mean that we must accept these conclusions. Up to now I have tried to show how seductive the arguments for these conclusions can be, and also how the conclusions themselves contribute to the illusion that materially adequate definitions of truth, à la Tarski, provide analyses of truth that capture the meaning, or intension, of ordinary truth predicates. It is now time to uncover the fallacies in these seductive arguments.

The argument that homophonic instances of Schema T_E are analytic is reconstructed in (19).

19. The argument that homophonic instances of Schema T_E are analytic
   a. If one is a competent speaker of English who knows what quotation, *iff*, and the phrase *is true in English* mean, then one will know that instances of *'s' is true in English iff p* are true, where an instance is the result of replacing the letter 's' with a "normal" (unparadoxical, nonindeterminate, non–context-sensitive) declarative sentence of English, and the letter 'p' with a sentence that means the same as the sentence that replaces 's'.
   b. So, if one is a competent speaker of English who knows what quotation, *iff*, and the phrase *is true in English* mean, then one will know that instances of *'s' is true in English iff s* are truths of English, where an instance is the result of replacing both occurrences of the letter 's' with a "normal" (unparadoxical, nonindeterminate, non–context-sensitive) declarative sentence of English.
   c. If, in addition, e is such a sentence, and one understands e, then one will accept, and know to be true, the instance of *'s' is true in English iff s* that results from substituting sentence e for each occurrence of the letter 's'.
   d. Since one understands this instance, accepts it, and believes (knows) it to be true, it follows by weak disquotation that one believes (knows) the proposition it expresses.
   e. Thus, homophonic instances of Schema T_E are analytic in the sense of A_1 and A_2.

Aside from worries about problematic sentences, such as those that are vague, ungrounded, or paradoxical, step (a) in the argument is all right. If one knows what '   ' *is true in English iff* means, then one knows that any instance of the schema *'s' is true in English iff p* which results from replacing 'p' with some sentence that means the same as the sentence that replaces 's' will express a truth (in one's language). The real problem with the argument occurs at steps (b) and (c). The case of (c) is particularly clear, and can be illustrated by comparing the sentences (20a) and (20b).

20a. 'Snow is white' is true in English iff snow is white.
   b. 'Snow is white' is true in my language iff snow is white.

Suppose that my language is English, but that I don't know that the word *English* refers to my language. I understand the word, and know a number of things about it – for example, that it designates a language spoken in England, North America, Australia, and New Zealand. However, I don't know that it designates the language I speak. This is clearly possible, just as it is possible for me to understand the name 'Japanese' without knowing that it is the language I hear on Channel 25, to understand the name 'Boston' without knowing that I am presently located in the city it designates, or to understand the name 'October' without knowing that it designates the current month. If I am in this situation, and I don't know that *English* designates my language, then my attitudes toward the sentence (20a) and (20b) may be different.

We may suppose that I know that (20b) is a true sentence of my language, that I accept it, and that I believe the proposition it expresses. Sentence (20a) is also a sentence of my language; moreover it is one I understand. Do I know that it expresses a truth? Do I accept it? Do I believe the proposition it expresses? Let us suppose not only that I understand the construction ' ' *is true in L iff*, but also that I know that both it and the word *English* are parts of the language designated by *English*, and that their meanings in that language are precisely what I take them to be (in my language).[19] We may then suppose that I accept (21a) and (21b).

21a. All sentences of the form *'s' is true in English iff s* which result from substituting a (proper) declarative sentence of English for each occurrence of '*s*' are truths of English.
 b. All sentences of the form *'s' is true in my language iff s* which result from substituting a (proper) declarative sentence of my language for each occurrence of '*s*' are truths of my language.

The sentence *Snow is white* is a sentence of my language which I understand and recognize as a sentence of my language. However, I may not realize that it is a sentence of the language designated by the name *English*. Thus, I may fail to assent to, or I may even dissent from, (22a), while accepting (22b).

22a. 'Snow is white' is a sentence of English.
 b. 'Snow is white' is a sentence of my language.

Because of this I may fail to accept, or I may even dissent from, (23a), while accepting (23b).

23a. *'Snow is white' is true in English iff snow is white* is a truth of English.
 b. *'Snow is white' is true in my language iff snow is white* is a truth of my language.

Next consider (24).

24. *'Snow is white' is true in English iff snow is white* is a truth of my language.

We may imagine that I accept this sentence iff I accept its disquotational version (20a) (which is just the italicized part of (24)). One reason for failing to accept these sentences has already been covered; although I may believe that snow is white, I may not realize that the words *Snow is white* constitute a sentence of the language designated by *English*. Moreover, even if I do realize this, there is another reason why I may fail to accept (24) and (20a); I may not believe that the sentence *Snow is white* means in the language designated by *English* what it means in my language. Indeed I may suspect that it means something in the language designated by *English* which has a truth value different from the truth value it has in my language. All of this goes to show that understanding and reflecting about homophonic instances of Schema T_E of the sort illustrated by (20a) does not guarantee that one has a sufficient basis for accepting them, believing them to be true, or believing the propositions they express. Hence, homophonic instances of Schema T_E of this sort are not analytic in the sense of A_1 or A_2.[20]

However, we still haven't determined whether (20b) is analytic; nor have we determined whether (20a) is a priori. Certainly if (20a) and (20b) express the same proposition, then (20a) is a priori if (20b) is.[21] But it is pretty clear that these sentences express different propositions.[22] The proposition expressed by my use of (20b) is a proposition about me. It can't be true in any world in which snow is white, but I don't use any language, or indeed in any world in which snow is white, but it is not the case that the sentence *Snow is white* is a true sentence of the language I use in that world. Since the same cannot be said for the proposition expressed by (20a), the two propositions are different. Thus, whether or not (20b) is analytic or a priori doesn't directly affect our question about whether (20a) is a priori.

Since one can understand the sentence (20a) while still requiring empirical information in order to accept it, or come to know it, the proposition expressed by (20a) is not knowable solely on the basis of entertaining it, as it is presented by (20a). Is it knowable solely on the basis of entertaining it, as presented in some other way – not, as we have seen, by (20b), but by (20c) perhaps?

20c. 'Snow is white' is true in that (demonstrating my language) iff snow is white.

This example is no doubt a little awkward, but no matter. The idea is to use the demonstrative to designate my language so that the left-hand side of the biconditional expresses a singular proposition about that language. If the left-hand side of (20a) expresses that same proposition, then the proposition expressed by (20a) will be knowable a priori, provided that the proposition expressed by the imagined use of (20c) is.

Presumably, however, the proposition expressed by (20c) is not a priori. To regard (20c) as a priori would be like regarding as a priori the proposition expressed by

25. He is less than 7 feet tall

on an occasion in which I am fully visible and *he* is used to refer to me. On such an occasion the empirical circumstances used, together with the sentence, to present a certain proposition include facts that allow one to verify that the proposition is true. The crucial point is that the empirical circumstances in question play a role not only in the presentation of the proposition, but also in justifying our belief in it. In this kind of case, entertaining the proposition and reflecting on it are not sufficient, without appeal to empirical facts, to come to know that it is true.

One indication of this comes from cases in which, after grasping the proposition, one comes to doubt that it is true, perhaps because one wrongly comes to suspect that one has been the victim of an optical illusion. To reassure oneself, in such a case, it will do no good simply to reflect on the proposition and reason about it; rather one needs to seek further empirical confirmation to determine its truth. The lesson here is that if we are careful to distinguish between the role of empirical facts in presenting a proposition and their role in justifying it, we will not be tempted to think that the proposition expressed by (25) on the occasion in question is a priori; or that the proposition expressed by (20c) is.

For these reasons, I believe that neither the proposition expressed by (20a) nor the proposition expressed by (20c) is a priori. The same point can be made about the proposition expressed by my use of (20b). This proposition talks about my language, which, I assume, is a public object; it is not an ideolect private to me, whose semantic properties are, by definition, just what I take them to be. Rather, my language is the language spoken by my community; I intend to speak that language, and I try to conform my usage to that of other speakers. When there is divergence between my usage and that of my community, I will typically take myself to be in error, not to be speaking a private ideolect of my own making. With this in mind, we can inquire into the status of T-sentences like (20b).

Suppose, for example, that I correctly believe that the word *podiatrist* means foot doctor, and that I correctly believe that *John is a podiatrist* is true in my language iff John is a foot doctor.

26. 'John is a podiatrist' is true in my language iff John is a foot doctor.

Suppose further that I am challenged, and I entertain the possibility that the word *podiatrist* really means something a little more specialized–say children's foot doctor. How would I go about checking to determine whether my original beliefs about my language were correct? In such a case, it would not be enough simply to introspect about the meanings I take sentences to have. Presumably, I would have to check the usage of other speakers of the language. Thus, to justify my beliefs, including my belief in (26), I would have to appeal to empirical facts. This suggests that the proposition expressed by (26), which is in fact also expressed by

27. 'John is a podiatrist' is true in my language iff John is a podiatrist

is not a priori. The same holds for (20b).[23]

Having decided that homophonic instances of Schema T_E like (20a) are not analytic, and that none of the examples in (20) is a priori, I will finish up by asking whether the T-sentences (20b) and (20c) are analytic. In order to answer this question we need a notion of analyticity that applies to sentences containing indexicals. One such notion, familiar from David Kaplan's work on the logic of demonstratives, is that a sentence is analytic if it expresses a truth in every context of every model of the language.[24] On Kaplan's conception, the meaning of a sentence is a function from contexts of utterance to propositions expressed by the sentence in those contexts. The idea is that a sentence is analytic if its meaning is sufficient to ensure that it always expresses a true proposition, even if the proposition expressed is not always necessary.

Whatever the merits of this characterization, (20b) and (20c) do not satisfy it. Sentence (20c) – *'Snow is white' is true in that iff snow is white* – is clearly not analytic in this sense, for there is nothing in its meaning that rules out contexts in which the demonstrative refers to any arbitrary language. In the case of (20b) we need to be careful what we mean by a context. Following Kaplan, we may take it to be any possible situation in which we might want to evaluate sentences relative to the point of view of an agent, at a time, place, and world. Now clearly there are contexts of this sort in which the agent does not speak English, and in which *Snow is white* is not a true sentence of the agent's language, even though snow is white in the world of the context. Since neither (20b) nor (20c) expresses a truth in every context, they are not analytic in the sense just considered.

Still, there is no denying that there is a sense in which T-sentences like (20b) are properly regarded as trivial by any English speaker who understands them. They are trivial in that anyone who uses and understands them on a given occasion will have reason to accept them and believe the propositions they express on that occasion. Such a speaker may even be taken to know these propositions (provided that we recognize that such knowledge is not a priori, but rather is based on knowledge of empirical facts about the use of expressions in one's linguistic community). If we like, we may dub this a kind of analyticity, one not far removed from the characterizations A_1 and A_2 considered earlier in connection with indexical-free sentences. Since T-sentences like (20b) are analytic in this sense, a person who understands his own language will know that which they express. Thus, the analyticity of T-sentences like (20b) guarantees that there is a sense in which competent speakers of a language know the truth conditions of the sentences of their own language.

It is important to realize, however, that it does not follow from this guarantee that one understands the sentences of one's language because one knows their truth conditions. In fact, for many sentences the situation may well be reversed. For example, one may know the truth conditions of the sentence *Electrons are subatomic particles* because one understands the sentence, knows what the truth predicate means, and therefore accepts (28).

28. 'Electrons are subatomic particles' is true in my language iff electrons are subatomic particles.

However, there need be no other interesting sense in which a person knows the truth conditions of the sentence mentioned in (28). If this is right, then although it may be true to say that understanding a language involves knowledge of its truth conditions, one cannot, in general, explain what understanding consists in by citing knowledge of truth conditions. One does not understand arbitrary sentences of one's language because one knows their truth conditions; rather one knows the truth conditions of many of these sentences because one understands them, and knows what it is to say that they are true.[25]

NOTES

1. in "Truth and Proof," *Scientific American*, June 1969, pp. 63–77, Tarski speaks at length of instances of Schema T as "partial definitions" that "explain the meaning" of the truth predicate as applied to particular object-language sentences. For example, on p. 64 he says the following:

   We begin with a simple problem. Consider a sentence in English whose meaning does not raise any doubts, say the sentence 'snow is white'. For brevity we denote this sentence by '*S*', so that '*S*' becomes the name of the sentence. We ask ourselves the question: What do we mean by saying that *S* is true or that it is false? The answer to this question is simple: in the spirit of Aristotelian explanation, by saying that *S* is true we mean simply that snow is white, and by saying that *S* is false we mean that snow is not white. By eliminating the symbol '*S*' we arrive at the following formulations:

   (1) Snow is white' is true if and only if snow is white
   (1′) 'Snow is white' is false if and only if snow is not white.

   Thus (1) and (1′) provide satisfactory explanations of the meaning of the terms 'true' and 'false' when these terms are referred to the sentence 'snow is white'. We can regard (1) and (1′) as partial definitions of the terms 'true' and 'false', in fact, as definitions of these terms with respect to a particular sentence. Notice that (1), as well as (1′), has the form prescribed for definitions by the rules of logic, namely the form of logical equivalence. It consists of two parts, the left and right side of the equivalence, combined by the connective 'if and only if'. The left side is the definiendum, the phrase whose meaning is explained by the definition; the right side is the definiens, the phrase that provides the explanation.

   If individual instances of Schema T are regarded as partial definitions that give the ordinary and recognized meaning of the word *true* when applied to particular sentences, then it is natural to regard the problem of giving a general definition that captures the ordinary and recognized meaning of the word when applied to an arbitrarily large or even infinite domain of sentences as that of finding an appropriate generalization that entails each partial definition. This is precisely how Tarski presents the problem. Thus, on p. 65 he says:

Thus our main problem is: can we establish an adequate use of the term 'true' for sentences in English and, if so, then by what methods? We can, of course, raise an analogous question for sentences in any other language. The problem will be solved completely if we manage to construct a general definition of truth that will be adequate in the sense that it will carry with it as logical consequences all the equivalences of the form (3) ['*p*' is true if and only if *p*]. If such a definition is accepted by English-speaking people, it will obviously establish an adequate use of the term 'true'.

2. Alfred Tarski, "The Semantic Conception of Truth and the Foundations of Semantics," *Philosophy and Phenomenological Research* 4, (1944): 341–75; reprinted in A. P. Martinich (ed.), *The Philosophy of Language*, (Oxford: Oxford University Press, 1985), pp. 48–71. The quoted remark occurs on p. 48 of the Martinich volume.
3. See sec. 13 (p. 57 in Martinich).
4. An instance of (1) is the result of replacing both occurrences of 's' with a name of a sentence of the object language L, and both occurrences of 'p' with a sentence of the metalanguage.
5. Italics indicate quotation; boldface italics indicate Quine's corner quotes.
6. A Tarskian definition of truth relative to a sequence (in our case an object) makes use, in clause (a) covering atomic formulas, of the notion of a predicate applying to an object and the notion of a singular term denoting an object relative to a sequence (object). These notions are not regarded by Tarski as primitives; rather they are given simple list definitions. In our case the definitions are as follows: A predicate P of L applies to an object o iff P = *W* and o is white. A singular term t of L denotes an object o relative to an object o′ iff t = *s* and o is snow, or t = *x* and o = o′. The parenthetical remarks in clause (a) of (3a) in the text are indications of this.
7. This is a formula of the metalanguage with one free variable 'S' ranging over object-language sentences.
8. For example, consider the instances (1_i) and (8_i) of schemata (1) and (8), respectively. (Here we take L to be a language spoken by some community, and we regard 'L' as a proper name of that language, in the way that 'English' is a name of English. We may also understand the conditionals as being material conditionals.)

1_i If *Ws* means in L that snow is black, then *Ws* is true in L iff snow is black.

8_i If *Ws* means in L that snow is black, then ([for all objects o, < *Ws*, o > is a member of a set T_s such that for all formulas G of L and objects o′, < G, o′ > is a member of T_s iff (a) G = *Ws* and snow is white or G = *Wx* and o′ is white; or (b) G = ***(~A)***, for some formula A of L, and < A, o′ > is not a member of T_s; or (c) G = ***(A & B)***, for some formulas A and B of L, and < A, o′ > and < B, o′ > are both members of T_s; or (d) G = ***(∃x A(x))***, for some formula A(x) of L, and there is an object o* such that <A(x), o* > is a member of T_s] iff snow is black.)

There are three points to notice about these instances. First, these are the instances that would be needed to draw the conclusion that *Ws* does not mean in *L* that snow is black from the T-sentences (5) and (6) respectively. Second, someone who knows what is meant by the ordinary predicates *true* and *means* is in a position to recognize the truth of (1_i), and to use it to draw conclusions about the meaning of *Ws*, whether or not that person understands any of the sentences of *L*; however, someone who does not understand any of the sentences of *L* is in no position to recognize the truth of (8_i). Third, the consequent of (8_i) is an a priori falsehood, since it is equivalent to the claim that snow is white iff snow is black. Thus, (8_i) as a whole could be an a priori truth, only if its antecedent were an a priori falsehood – i.e., only if the negation of the antecedent, *'Ws' does not mean in L that snow is black,* were knowable a priori. But this, I take it, is not a priori, just as statements about what English sentences mean, or do not mean, are not a priori. (The question of whether claims like *'Snow is white' is true in English iff snow is white* are a priori is discussed further below. Much of that discussion carries over to related claims like *'Snow is white' means in English that snow is white.*)

If this is right, then there are many instances of schema (8) that are neither trivial nor a priori. Of course, some instances will be both, but only because the consequents of those instances are–as when *Ws* replaces 's' and *snow is white* replaces '*p*' in (8). However, even in these cases the relevant instances are of no use in drawing conclusions about meaning. Moreover, the contrast with (1) remains intact. In the case of (1) arbitrary instances are trivial and a priori–no matter what replaces '*s*' and '*p*'. This is not so with (8).

9. A version of this argument is given in sec. III of my "What Is a Theory of Truth?" *Journal of Philosophy* 81 (1984): 411–29.
10. *Word and Object* (Cambridge Mass.: MIT Press, 1960), p. 58.
11. La Salle, Ill.: Open Court, 1974.
12. Ibid., p. 78.
13. Ibid., p. 79.
14. Of course, if we say that homophonic instances of Schema T are analytic without being necessary, whereas the T-sentences derived from Tarski's stipulative definition of truth are necessary, then we will weaken the appeal of premise (iii) in the argument given in the text for the (erroneous) conclusion that Tarski's definition captures the meaning of the ordinary truth predicate, when restricted to a well-behaved object language. Nevertheless, the question of whether homophonic instances of Schema T are analytic in the sense of A_1 or A_2 remains interesting in its own right.
15. Saul Kripke, "A Puzzle about Belief," in A. Margalit (ed.), (Dordrecht: Reidel, 1979), pp. 239–83, reprinted in N. Salmon and S. Soames, (eds.), *Propositions and Attitudes* (Oxford: Oxford University Press, 1988), pp. 102–48.
16. "A Puzzle about Belief," at p. 134 of Salmon and Soames *Propositions and Attitudes.*
17. "How to Become a Millian Heir," *Noûs* 23, (1989): 211–20, at pp. 216–17. See also "A Millian Heir Rejects the Wages of *Sinn,*" in C. A. Anderson and J. Owens, (eds.), *Propositional Attitudes: the Role of Content in Logic, Language, and Mind* (Stanford, Calif:, CSLI), 1990, pp. 215–47.
18. "Understanding synonyms without knowing that they are synonymous," *Analysis* 52 (1992): 224–8.
19. The need for these extra conditions is shown by the following considerations. First, if I did not know that '   ' *is true in English iff* were a construction in the language designated by the word *English,* then I would not know that instances of *'s' is true in English iff s* were themselves sentences of the language designated by that word. Second, if I did know that these were sentences of the language designated by *English,* but if I had the wrong idea about what '   ' *is true in English iff* meant in that language–for example if I wrongly thought that it meant what '   ' *is false in English iff* means in my language–then I would not know that instances of *'s' is true in English iff s* were truths of the language designated by *English.* In either of these cases I would not accept sentence (21a). Since no other sentence I am guaranteed to accept expresses the same proposition as (21a), I might also fail to believe, or know, that proposition. This is the problem mentioned with step (b) of the argument in (19). That step, which did not include the

extra conditions noted here, is false, as stated. I am indebted to Doug Cannon for a discussion of these issues.

20. I have occasionally been asked whether this conclusion could be avoided if sentences were thought of not simply as syntactic objects (sequences of shape or sound types), but rather as syntactic objects together with their interpretations. On this conception, the English sentence *Snow is white* is thought of as a pair, the first member of which is a sequence of syntactic elements, and the second member of which is the proposition that snow is white. It is sometimes further suggested that quotation mark names of sentences, like *'Snow is white'*, name the pair consisting of the sequence of syntactic elements, together with the proposition it expresses in English. On this conception, the truth predicate that occurs in instances of Schema T is not relational (it does not hold between sentences and languages), nor is it indexed to a particular language. Rather, a sentence like *'Snow is white' is true in English* is understood as *'Snow is white' is a sentence of English which is true*, with the truth predicate expressing a nonrelational property of "interpreted sentences" (propositions really). Now, I do not accept this account of quotation, or of what sentences are. However, it is worth noting that accepting the account would not change our results. On the conception sketched, knowing that 'Snow is white' is a sentence of English would involve knowing that the pair consisting of a certain syntactic object and the proposition that snow is white was a sentence of English, which in turn would require knowing that that syntactic object expressed that proposition in English. This means that, for all the reasons given in the text, understanding the T-sentences (when construed in this way) would not guarantee that one accepted them, or believed the propositions they expressed.
21. Propositions are, of course, the primary bearers of a prioricity; however, we may speak of sentences, or even particular uses of sentences, as being a priori in the derivative sense of expressing a proposition which is a priori.
22. More precisely, (20a) expresses a different proposition from any individual's – in this case my – use of (20b).
23. Could this conclusion be resisted by granting that the proposition which is expressed by both (26) and (27) is not knowable independent of experience, when it is presented by (26), but is knowable without appeal to experience, when it is presented by (27), and so is a priori? I don't think so. I concede that, no matter what doubts I might have about whether I have properly understood the word *podiatrist*, I will believe, perhaps even know, that sentence (27) expresses a true proposition in my language. But in order for me to know that proposition to be true, as it is presented to me by sentence (27), I must know that the meaning that I take *podiatrist* to have is the proper one, and hence that the proposition I take the right side of the biconditional (27) to express is the right one, as determined by the rules of my language. Since this can be verified only by appeal to empirical facts about how the word *podiatrist* is used in my linguistic community, my knowledge of that which is expressed by (27) is a posteriori.
24. David Kaplan "Demonstratives," in J. Almog, J. Perry, and H. Wettstein (eds.), *Themes from Kaplan* (New York: Oxford University Press, 1989), pp. 481–564.
25. A version of this paper was read at the Colloquium on Truth and Truth Definitions sponsored by the Boston Colloquium for the Philosophy of Science, held at Boston University, October 23–24, 1992. I am indebted to Ali Akhtar Kazmi and Doug Cannon for useful discussions of it.

For EU product safety concerns, contact us at Calle de José Abascal, 56–1°,
28003 Madrid, Spain or eugpsr@cambridge.org.

www.ingramcontent.com/pod-product-compliance
Lightning Source LLC
LaVergne TN
LVHW090314210926
848480LV00001B/238

* 9 7 8 0 5 2 1 1 0 0 5 7 1 *